THE COSMIC ARCHITECTURE

Understanding the Universe, Our Place, and Our Purpose

RISHABH KUMAR JAIN

THE COSMIC ARCHITECTURE

This book presents a speculative philosophical and systems-based thought experiment. References to physics, biology, computing, engineering, religion, and philosophy are used for explanation, comparison, and conceptual exploration. Speculative interpretations presented in this work should not be understood as established scientific conclusions.

First Edition, 2026
ISBN: 978-1-7648938-1-7

Written by Rishabh Kumar Jain

ACKNOWLEDGEMENTS

In loving memory of my parents, whose love, values, and sacrifices laid the foundation for everything I have achieved; and to my siblings, nieces, nephew, and in-laws, whose love and encouragement continue to support me. This book is dedicated to you all.

This book grew from a lifelong curiosity about some of humanity's oldest questions: Why does the universe exist? What is consciousness? Can science and spirituality illuminate one another rather than compete?

My perspective has been shaped by two very different worlds. Professionally, I have spent years working in software engineering, cloud computing, distributed systems, and artificial intelligence. My perspective has also been shaped by the study and practice of Jain philosophy. During *Paryushan*, the most sacred period in the Jain calendar, I *observed* ten consecutive days of water-only fasting for two successive years. Those experiences provided time for deep reflection and reinforced the values of discipline, compassion, non-violence, and intellectual humility that have influenced many of the ideas explored in this book.

The Cosmic Architecture is not presented as established science or religious doctrine. It is an invitation to explore whether the language of modern systems engineering can provide a fresh way of thinking about reality, while respecting both scientific evidence and the wisdom found in philosophical traditions.

I am deeply grateful to my family, teachers, colleagues, and the many scientists, philosophers, engineers, and spiritual thinkers whose work inspired the questions explored in these pages. I also thank every reader who approaches this book with both curiosity and critical thought. Whether you ultimately agree with its ideas or not, I hope it encourages you to ask deeper questions about the universe and our place within it.

A NOTE ON SCIENCE, SPECULATION, AND PURPOSE

This volume spans multiple disciplines: established astrophysics, quantum field theory, data engineering, systems architecture, philosophy, and ancient spiritual traditions. Because it integrates fields that rarely sit beside one another, establishing a clear perspective from the outset is essential.

Throughout this book, whenever established science is discussed—such as Einstein's General Relativity, the invariant speed of light (c), cometary prebiotic chemistry, or evolutionary biology—the physical mechanisms are presented accurately as standard science understands them today. Similarly, when modern computing concepts are introduced—such as relational databases, data warehousing, vector embeddings, or multi-tenant cloud platforms—they reflect real-world systems engineering principles used in technology today.

However, the overarching synthesis presented in this work—what I call the **Cosmic Architecture**—is an **ambitious philosophical model**, a **speculative conceptual framework**, and a **structured thought experiment**. It asks a fundamental question:

What if the observable processes of physical nature are not cold, isolated accidents, but the visible output of an unimaginably advanced, high-dimensional system designed to host, evaluate, and elevate conscious life?

Disclaimer: Rather than repeating *disclaimers* in every chapter, the reader is invited to approach this work as a journey of **disciplined imagination**. This book does not claim to present new, laboratory-tested physics equations, nor does it demand uncritical religious faith. Instead, it uses established scientific facts and modern systems concepts as analogies—conceptual scaffolding for exploring a logically structured "what if" model of reality. Whenever the book moves beyond evidence into interpretation, that movement should be understood as philosophical speculation rather than scientific proof.

READER'S NOTE

This volume is designed as a modular master-reference. If you are driven by physics and software, Parts I, II, and V lay the technical foundation. If you are seeking deep philosophical and moral inquiry, Parts III, VI, and VIII explore the human journey. Read linearly or jump directly to the modules that call to you.

The purpose of *The Cosmic Architecture* is to offer a bold, integrated perspective on life, consciousness, morality, and the multiverse. It addresses deep existential questions—such as why the universe is so vast, why human choices carry weight, why suffering occurs, and where consciousness goes upon physical shutdown—by looking through the lens of modern systems engineering.

The reader is invited to approach these pages with both an open mind and a critical, disciplined perspective. This book is neither a dry technical handbook nor a piece of fantasy fiction; it sits in the compelling space between hard science, systems philosophy, and speculative cosmology.

Where science provides empirical facts, we accept them completely. Where science reaches its current observational limits, *The Cosmic Architecture* steps forward with a structured, imaginative model—inviting you to explore a universe where physical laws, human free agency, and cosmic purpose operate as a single, magnificent, and unified whole.

PART I — HOW TO READ THE ARCHITECTURE

Looking Back:

In the opening notes, we established the guiding principles of this volume. Physical nature is approached first through empirical science, while the overarching framework of ***The Cosmic Architecture*** remains a disciplined thought experiment and speculative model. We also established an important rule for what follows: terms borrowed from software engineering should not be interpreted literally. There are no claims of cosmic server racks, Git servers, or source-code folders hidden somewhere in space. Instead, familiar technologies provide a conceptual ladder for asking a much larger question: *Can the principles humans use to understand complex systems help us imagine architectures of reality that may lie far beyond our present engineering vocabulary?*

Moving Forward:

Before exploring quantum physics, biology, consciousness, or human destiny through this systems lens, we first need to learn how to read its architectural language. **Part I** introduces repositories, commits, branches, release tags, nodes, networks, hierarchies, vectors, and high-dimensional state-spaces—but then pushes beyond these familiar concepts.

We will ask what a cosmic "repository" could mean if no central source code were ever stored in one discoverable location; how an architecture might instead be distributed across deeper structures of reality; how a universe could be represented as many sectors with different states and relationships; and how apparently distant sectors could be compared within an enormous state-space without implying millions of literal physical dimensions. We will also distinguish **similarity from connectivity**—why two cosmic sectors could be close in state yet not necessarily connected.

These ideas are not presented as established mechanisms of nature. They are the **conceptual scaffolding** upon which the later Cosmic Architecture will be built.

By the end of Part I, terms such as *repository, branch, node, vector, sector,* and *dimension* will no longer mean quite what they did at the beginning. They will become a shared architectural vocabulary for exploring a possibility far larger than today's technology: **that our most advanced engineering concepts may be useful not because the cosmos literally uses them, but because they provide primitive human approximations of organizational principles that could be vastly more sophisticated than anything we currently know how to build.**

Chapter 1

Cosmic Patterns Before Code

The **Cosmic Architecture** begins with a fundamental realization: *before we can explore whether **reality** possesses a deeper structural design, we must first learn how to see the universe through the **lens of systems**. A system is simply a group of parts working together as a connected whole. A living cell, an ecosystem, a galaxy, a climate, and a human civilization are all systems—and in every case, the relationships between the parts matter just as much as the parts themselves.*

This matters because many later questions in the book are about shared dependence. A decision by a central bank can affect millions of borrowers. A solar storm can affect satellites, power grids and communication systems. A small change in a common software tool can affect many applications that depend on it. In each case, the local events look different, but they are connected by something deeper upstream.

The book uses this systems idea to ask a bigger question. If reality has deeper layers than we normally see, could changes move through shared structure rather than only through relationships obvious to us? The *Cosmic Architecture* proposes that possibility as part of its model. Science has not established such a hidden layer, but known science also describes only the interactions it can currently observe and measure. The purpose here is to explore what a larger system might look like if such layers exist.

A variable is something that can change, such as temperature, population, speed or blood pressure. A setting is an adjustable value that helps a model operate in a certain way. A constant is something treated as fixed inside a theory. These words may sound similar in everyday language, but science uses them more carefully. Changing a value inside a universe is not the same as imagining a universe with different basic constants.

When people talk about fine-tuning, it can be tempting to imagine the universe as if it has a hidden settings screen. That image can help beginners, but it is only an analogy. The real questions are about physical constants,

1. Clearing Up the Confusion: Databases vs. Source Repositories

Before we dive into how the universe stores its master blueprints, we need to make a very clear distinction between two standard technology terms that often get mixed up:

- **The Cosmic Database & Ledger (Chapters 5, 15, and 19):** This is the system that handles **active, daily runtime data**. It records your personal choices, your individual soul state vector, your karmic updates, and your daily life experiences. Think of it like a bank's transactional database—it tracks your account balance and every transaction you make while the system is running.
- **The Cosmic Source Repository:** This system stores the **master software code and foundational blueprints**. It holds the master physical equations, the strength of gravity, the speed of light, chemical interaction tables, and the master templates for biological species. Think of it like the original blueprint plans for the bank's entire computer network and building structure.

A database tracks **what happens inside an active application**; a source repository stores the underlying blueprints used to build and spin up the application in the first place.

2. The Mechanics of Version Control: Git in Plain English

To understand how a Cosmic Repository works, imagine writing a complex 500-page book over three years on your laptop.

If you save every daily edit over the same file (MyBook.docx), a single accidental deletion or bad revision can ruin months of hard work forever. Your document only knows its current state; it is completely blind to the journey that brought it there.

To solve this, developers created version control tools like **Git**. Instead of relying on a single overwritten file, Git records a structured history of changes made to a project.

Four simple terms explain this logic:

1. **The Repository (Repo):** The main master container holding the complete project, its entire historical code baseline, and all version tags.
2. **The Commit:** A *snapshot* saved at a specific moment in time. In normal use, commits are treated as stable historical records. Git history can be deliberately rewritten, but doing so changes commit identities and can disrupt the integrity of a shared history. New commits normally build on the baseline before them.
3. **The Branch:** A parallel, independent line of development created to test new ideas, run experiments, or build new features without corrupting the main codebase. Developers can experiment freely inside a branch; if the experiment succeeds, it can be merged back into the main project. If it fails, the branch can be isolated or discarded without crashing the main system.
4. **The Tag:** A human-readable label attached to a specific commit to mark a stable milestone or release version (for example, labelling a stable build as Version 1.0).

While these terms come from software engineering, the underlying logic is universal. They show how a vastly complex system can allow local movement, testing, and experimentation while keeping its historical memory completely safe and intact.

3. Cosmic Branching: How Parallel Universes Spin Up

Applied to cosmology, this analogy suggests a way to imagine universes as developmental branches emerging from deeper shared structure. This is not a claim that parallel universes literally run on Git or that nature contains software repositories. It is a modelling device for asking how inheritance, variation and preservation might operate if reality has layers beyond the observable universe.

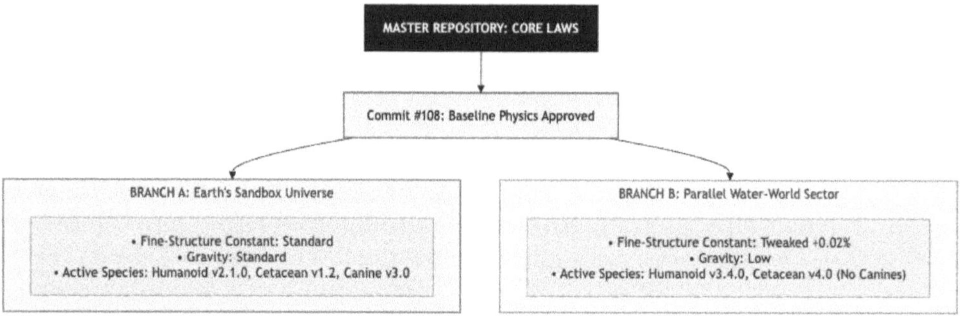

Real-World Analogy: Software Beta Testing

Imagine an entertainment company building a massive online multiplayer video game. Before releasing an update to millions of active players on the main servers (Production), the developers create an isolated testing branch (Sandbox Server). In that testing branch, they adjust gravity settings, test new character models, and see how the world behaves. If the testing branch works well, the improved character blueprints are saved back to the master library.

Within the *Cosmic Architecture*, this software example becomes a useful metaphor for cosmic cycles:

- **Spawning a New Multiverse Branch:** In the metaphor, a new universe cycle may be imagined as inheriting selected structural lessons from earlier cycles rather than beginning from a completely blank slate. This should be read as speculative modelling, not as a literal claim about cosmic version control.
- **Independent Experimentation:** That new universe runs its course for billions of years. Stars form, chemistry develops, and biological life evolves inside that branch.
- **Merging the Lessons:** When a universe reaches its physical conclusion, the model imagines that some compressed lesson of its history could persist in a deeper framework. This is a philosophical way to think about preservation, not a demonstrated archive beyond the observable universe.

4. Semantic Versioning in Layman's Terms

The *Cosmic Architecture* models the organization and tracking of species templates across different universe branches through a system analogous to **Semantic Versioning (SemVer). What is Semantic Versioning?**

Think of a smartphone operating system update represented by three numbers separated by dots (for example, **v3.2.1**):

- **MAJOR Version (3.x.x):** A massive, fundamental structural overhaul. Upgrading from v2 to v3 breaks compatibility with old systems because the baseline architecture changed completely (like transitioning from a 2D physical body to a 4D light-based vessel).
- **MINOR Version (x.2.x):** A significant feature addition or evolutionary enhancement that adds powerful new capabilities while working smoothly with the existing foundation (like adding opposable thumbs, expanding sensory vision, or developing vocal cords on a mammal).
- **PATCH Version (x.x.1):** A small bug fix, micro-adjustment, or environmental adaptation that cleans up minor errors without changing the core functionality (like adjusting skin pigmentation for high UV radiation or tuning metabolic efficiency for cold climates).

5. Species Versioning: Universal Templates vs. Localized Codebases

Within the model, if we imagine species repositories across a multiverse, a clear distinction emerges between the **Universal Humanoid Template** and **Localized Animal & Ecosystem Blueprints**.

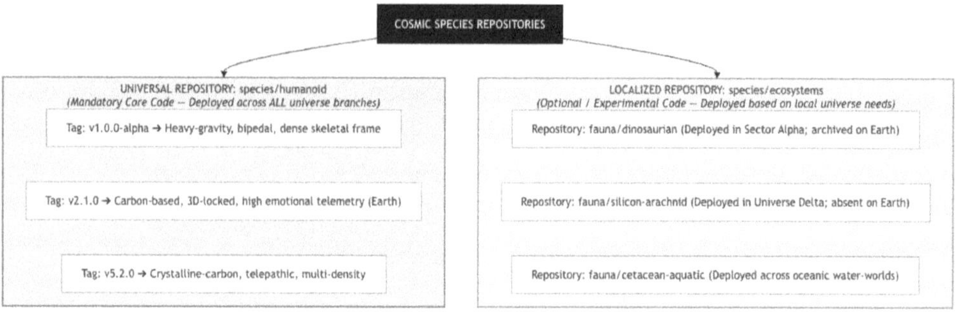

A. The Humanoid Template: The Universal Standard

Within the Cosmic Architecture model, the humanoid form—a bipedal body with upright balance, two precision hands, land-based mobility, vocal communication, and high-capacity neural architecture—is treated as one recurring vessel through which advanced moral agency could emerge.

The model therefore imagines that humanoid-like forms could recur across multiple universe domains where moral choice, technological civilization, and conscious stewardship emerge. This is a speculative design principle of the framework, not a biological claim that every possible universe must contain humanoids. Within the analogy, the version tag could vary dramatically with a universe's developmental stage:

Example 1: Developing Universe Sandbox (humanoid: v1.1.0)

In a young, rough universe imagined as running early physical code, a humanoid species might be tagged with a primitive version. It might have heavy bone density to withstand volatile gravity, thick skin for UV protection, and simple vocal syntax. Even so, it would retain the core bipedal blueprint needed to build tools, make moral choices, and begin the journey of civilization.

Example 2: Current Earth Sandbox (humanoid: v2.1.0)

Our current biological platform—a delicate balance of physical vulnerability, 3D spatial locking, high-calorie brain consumption, and deep emotional telemetry.

Example 3: Advanced Multiverse Sector (humanoid: v5.2.0)

In an older, highly evolved universe branch imagined by the model, a humanoid species might be represented as an advanced release, with expanded intuitive perception, broader sensory bandwidth, greatly reduced biological disease, and possible interaction with higher-dimensional aspects of reality.

B. Animal Species: Localized, Experimental, and Universe-Specific Code

Unlike the humanoid template, animal and plant species do not need to exist in every universe.

Within the Cosmic Architecture model, animals can be interpreted as serving ecological, terraforming, and emotional-development functions within local planetary sandboxes (as explored in Chapters 22 and 35). Because hypothetical planetary environments could vary widely across a multiverse, the analogy represents animal species as modular, localized repositories that might be deployed, omitted, or heavily customized:

Example 1: Universe-Specific Animal Blueprints

A young universe with dense, methane-rich atmospheres might deploy giant, silicon-based insectoid species (fauna/silicon-arachnid: v1.0.0) to break down mineral crusts. That exact animal species may never exist anywhere in our universe.

Example 2: Deprecated Animal Branches

In the framework's software analogy, Earth's long dinosaur era can be pictured as a major ecological phase in which large animals

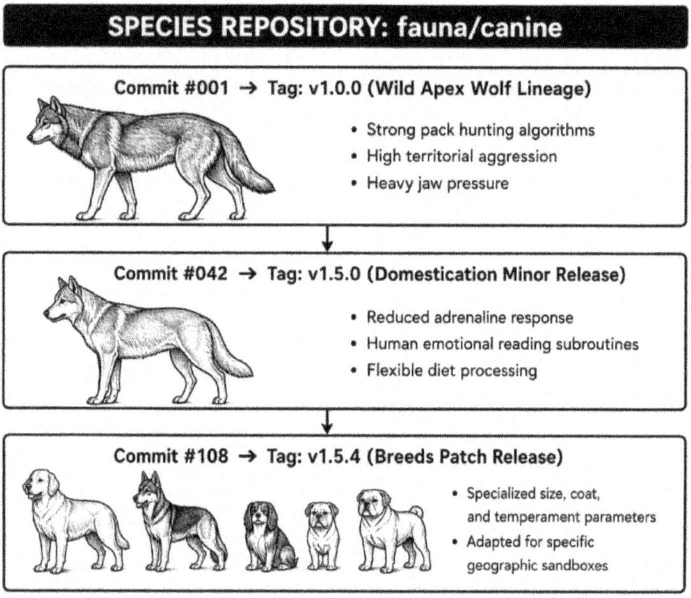

shaped vegetation, soils, nutrient cycles, and food webs over immense spans of time. Their later extinction can then be represented metaphorically as the retirement of one dominant biological codebase. In a hypothetical parallel universe, the model can imagine dinosaur-like lineages continuing to evolve alongside other intelligent species.

6. How Species Versioning Works Within Cosmic Architecture

To see how semantic versioning works across deep time, look at how a single species line updates its code through commits, minor releases, and patches:

A Speculative Real-World Example: The Canine Upgrade Pipeline

1. **Major Release (v1.0.0):** The raw ancestral wolf template is deployed to wild biospheres. It features high aggression, strong pack

coordination, and physical independence designed purely for wild survival.

2. **Minor Release (v1.5.0):** As humanoid nodes establish permanent settlements, the canine codebase receives a minor release. The system dials down wild flight-or-fight responses and enables new emotional reading subroutines. The canine node can now read human facial expressions, process human vocal tones, and form cross-species emotional bonds.
3. **Patch Releases (v1.5.4):** Local environmental conditions trigger minor patches—adjusting fur density for Arctic cold or tuning body size for specific tasks—without changing the core domesticated codebase.

7. Beyond Git: The Limits of the Repository Analogy

Beyond the Primitive Repository

To make these ideas visible, we began with familiar human technology: Git repositories, commits, branches, and release tags. These concepts are useful because they show how complex systems preserve history, manage controlled variation, and release stable configurations.

A reader familiar with DevOps, platform engineering, or cybersecurity may immediately raise an important objection: **If the architecture of reality depended upon one discoverable Master Repository, would that not create an enormous security risk and single point of failure? What would prevent corruption—or an advanced internal civilisation from eventually discovering and copying the source architecture itself?**

That objection reveals the limit of the Git analogy.

Human beings normally preserve source code by **storing it somewhere**. Even when large organizations use many repositories, access controls, backups, encryption, and separate teams, the underlying information still exists on identifiable physical infrastructure. If sufficiently privileged access is compromised, substantial portions of the architecture may potentially be exposed.

Within the speculative *Cosmic Architecture*, a creation-scale system need not preserve its most sensitive architecture in this way at all.

A Blueprint That Is Never Stored Whole

Instead of one permanent Master Repository, imagine the underlying architecture being **fragmented across an enormous distributed cosmic substrate**. Different pieces of structural information could exist across different regions, fields, or higher-order relationships, potentially extending across multiple universes.

No ordinary location would contain the complete blueprint. More importantly, the complete "source code" need not normally exist anywhere as a readable object.

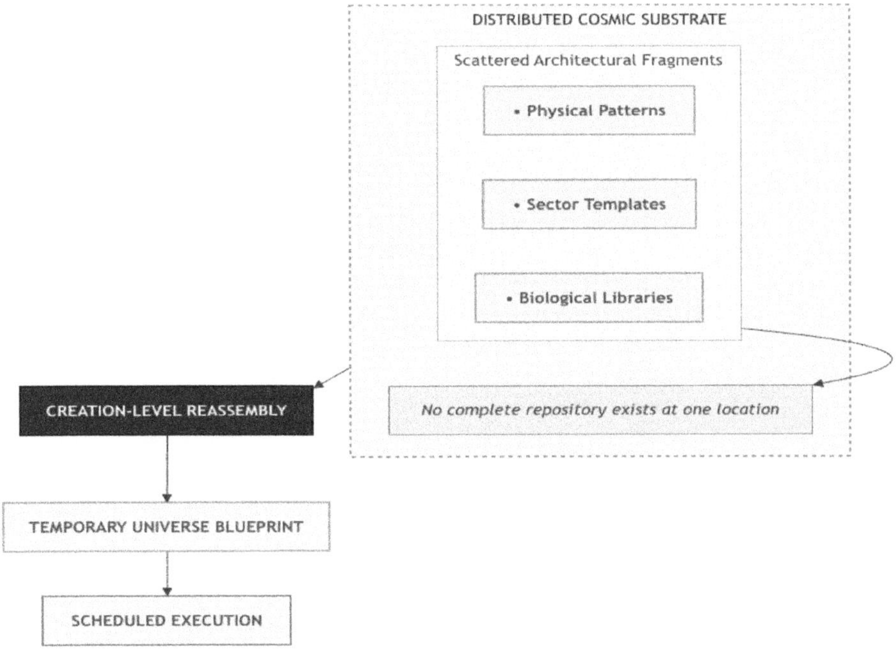

A. Distributed Substrate Memory

The "repository" should therefore not be imagined as a hidden file system somewhere in deep space. Within the model, fragments of foundational information could instead be distributed across deeper structures of spacetime, fields, or higher-order relationships in forms unknown to present science.

One fragment alone would reveal little. Thousands or vastly more such fragments, their relationships, their correct ordering, and the means of interpreting them would be required to reconstruct the larger architecture.

The repository therefore exists primarily as a **potential for authorized reconstruction**, rather than as a permanently assembled source tree.

B. Domain-Isolated Architectural Components

Human enterprises already separate sensitive systems according to function and access. The Cosmic Architecture extends this principle far beyond conventional repositories.

Using familiar notation only as shorthand, different fragments might contribute to logical domains such as:

core-laws/spacetime — fundamental physical constraints
cosmic-structure/galaxies — large-scale cosmic formation
environment/biospheres — planetary and ecological conditions
species/humanoid — one speculative conscious-life template
species/fauna — localized biological variations

These are not proposed as literal folders. They represent different functional components of an architecture whose underlying information could be distributed rather than stored together.

C. Neural–Cosmic Reassembly

The most important difference from human software engineering is **who—or what—performs the assembly**.

A hypothetical creation civilisation trillions of years beyond humanity might no longer separate biological intelligence, computation, and cosmic information systems in the way we do. Within this thought experiment, its creators could possess a direct interface between their cognitive architecture and the deeper informational substrate of reality.

A creator would therefore not need to log into a repository, download source code, hold a design meeting, or manually merge branches. The required fragments could be identified, interpreted, combined, and temporarily instantiated through a creation-level cognitive system.

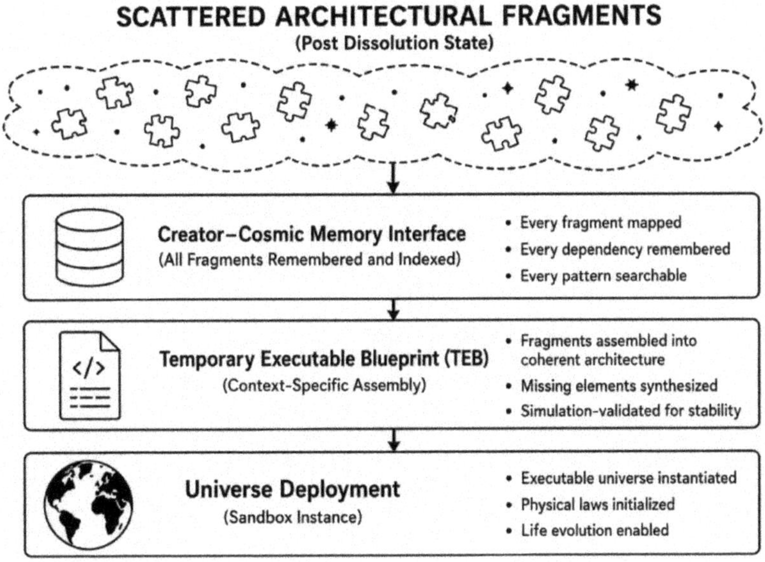

The words *neural* and *cognitive* should not imply an ordinary human brain scaled upward. They are placeholders for a form of intelligence and information processing that could be as far beyond the human nervous system as our present technology is beyond the earliest biological signalling networks.

D. Ephemeral Blueprint Synthesis

When a new universe such as **Universe Beta** is initialized, the model therefore does not require a static copy of Universe Alpha.

A temporary architecture could instead be reconstructed from a common foundational pattern together with selected inherited configurations, local variations, and **sector relationships**.

A sector relationship means that a particular region or domain of Universe Beta could be architecturally associated with a compatible sector elsewhere in the wider multiverse. One sector might correspond to a structural pattern associated with Universe Alpha, another with Universe Gamma, and another with Universe Z.

These sectors need not be physically adjacent in ordinary three-dimensional space. Their correspondence belongs to the model's wider relational architecture.

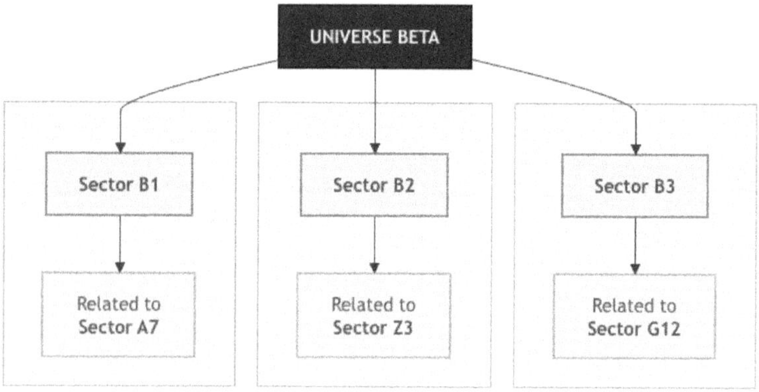

Similarity alone does not create these connections. As later chapters explain, **state-space proximity describes similarity, while the relationship layer determines which sectors are actually linked within the model**.

Universe Beta is therefore not a copy-and-paste deployment. It is a **unique composed instance** assembled from a common foundation and its own authorized combination of configurations and sector relationships.

E. Branching Across Cosmic Time

The assembled blueprint also need not execute everything simultaneously.

Its equivalent of "branches" could represent configurations authorized to become active at different stages of cosmic development:

Base Physics → Sector Formation → Galaxies → Stars → Planets → Chemistry → Biospheres → Flora and Fauna → Conscious Life

A branch scheduled for biological development, for example, need not become relevant while the universe is still forming its first stars.

In this sense, a cosmic release could unfold over billions of years. The actual mechanisms observed within each stage remain those studied by physics, cosmology, chemistry, geology, and evolutionary biology; **branching, scheduling, and deployment are the systems language used by this model to represent their architectural sequence.**

F. The Root-Access Barrier

This produces a very different security model from human software engineering.

An advanced civilisation inside a universe might eventually understand extraordinary portions of its local runtime. It might manipulate genetics, engineer stars, or discover physical phenomena far beyond present human knowledge.

But **understanding the runtime is not equivalent to possessing the architecture that instantiated it**.

There is no single Master Repository to steal. No complete source tree is permanently exposed. Individual fragments reveal neither the complete architecture nor necessarily how other fragments are located, interpreted, ordered, authorized, or assembled.

Most importantly, within this speculative model, reconstruction depends upon a creation-level cognitive interface unavailable to ordinary internal civilizations. Possessing fragments would therefore not automatically provide the capability to rebuild the whole.

Once the required configuration has served its purpose, the temporary assembled blueprint need not remain as a persistent repository. Its components remain distributed within the deeper architecture until some future authorized reconstruction requires them again.

The security principle is therefore not merely **better encryption**. It is more fundamental:

the complete source architecture is never persistently stored in a form available to the systems running inside it.

Git therefore remains a useful conceptual ladder. A *repository* represents preserved architecture; a *commit* a stable configuration; a *branch*-controlled variation; a *release tag* an identifiable state; and a *deployment* the realization of architecture across time.

But the Cosmic Architecture ultimately moves beyond Git. **Human beings store source code and then protect access to it. In this speculative model, creation-level intelligence protects the deepest architecture by never exposing the complete source as a repository in the first place.**

Chapter 3
State-Space Geometry: Mapping Similarity Beyond Physical Distance

A database is simply an organised way of storing information so that it can be found again later. A library catalogue is a familiar example. Modern databases may contain billions of records, but their basic purpose remains simple: organise information so that the right information can be retrieved when it is needed.

Modern artificial intelligence has popularized an important tool known as a vector database. Unlike a traditional database, which searches mainly by names, keywords or locations, a vector database organises information according to its overall state or meaning.

Every object is represented by a mathematical description called a **vector**. The vector is not the object itself. It is simply a way of describing the object's current state.

This distinction is important because it introduces one of the fundamental ideas used throughout the *Cosmic Architecture.*

A **node** is the thing that exists.

A **vector** describes the current state of that node.

For a human being, the state might include age, health, memories, knowledge, personality, relationships and countless other characteristics. For a planet, the state could include its mass, atmosphere, chemistry, temperature, water content and stage of development. For a galaxy, it might describe the number of stars, its structure, age and chemical composition. If we imagine an entire universe as a node, its state might include its physical laws, expansion rate, age and other properties that we have not yet discovered.

Artificial intelligence uses these mathematical descriptions to search for **nearest neighbours**. This does not mean the nearest object in physical space. It means the nearest object whose overall state is most similar.

Imagine the world's largest natural history museum containing billions of exhibits. One way to organise the museum would be alphabetically. Another would be to organise everything according to similarity. Birds would naturally appear together. Marine animals would occupy another section. Flowering plants would form another. Different kinds of rocks would create their own neighbourhood. Visitors looking at one exhibit would automatically discover many other related exhibits nearby, even though they have different names and originate from different parts of the world.

Artificial intelligence organises information in much the same way. Similar objects naturally form neighbourhoods inside a high-dimensional mathematical space.

The *Cosmic Architecture* extends this idea much further.

Suppose every node possesses its own state. Human beings, animals, forests, planets, stars, galaxies and perhaps even universes could all occupy positions within a vastly richer state-space. Physical distance would remain real, but another form of nearness might also exist—a nearness based upon shared characteristics rather than shared location.

Two planets separated by millions of light-years might nevertheless occupy neighbouring regions because both possess liquid water, rocky surfaces and developing biological life. Two civilisations might become neighbours because they have reached similar stages of scientific or moral development. Even two universes could occupy neighbouring states if they shared similar physical laws or histories despite being unimaginably distant in ordinary space.

This does not mean the universe literally operates like a vector database, nor that every spiritual or moral question can be reduced to mathematics. The analogy serves a narrower purpose: it shows how similarity can be meaningful even when physical distance is large. Later chapters use "vector," "nearest neighbour," and "state-space" as shorthand for this idea rather than as proof of a hidden cosmic database.

Why Universes Can Also Be Nearest Neighbours

The reason this framework extends nearest-neighbour relationships beyond planets to entire universes is important. If a conscious human node reaches the end of its physical life on Earth, the Cosmic Architecture does not necessarily require its next compatible embodiment to occur on Earth—or even within the same universe. In the speculative model developed later in this book, a wider multiverse may contain Earth-like planetary environments at different stages of development.

A nearest-neighbour process could therefore match the node's accumulated state with the most compatible environment available across this larger state-space. That environment might be broadly similar to present-day Earth, less developed, or substantially more advanced. Here, **"nearest" means closest in relevant conditions and compatibility, not closest in physical distance.**

The next chapter builds upon this idea. Once every object is understood as a node with its own state, the next question naturally follows: **How are those nodes connected?**

Chapter 4

Nodes, Networks and Hierarchy

1. Beyond Generic Units: Purpose-Driven System Nodes

The previous chapter introduced one of the foundational ideas of *The Cosmic Architecture*: every entity in existence holds a specific state that can be mapped within a broader system. In modern systems engineering, any distinct, identifiable participant inside a larger framework is referred to as a **node**.

A node is not defined by its physical size. It can be as small as a single biological cell or as vast as an entire universe. What changes across these scales is not the fundamental definition of an entity, but its operational scope, its capacity for processing, and its unique purpose within the master system.

When we view *reality* through an architectural lens, treating every entity as a generic "node" can obscure what makes different forms of existence unique. Within this speculative framework, it is useful to categorize entities not merely by their existence, but by the functional roles they may play within a broader network:

- **A Human Being as a Conscious Execution Unit:** Human beings are not merely passive physical objects. Within the model, we are represented as high-level processing nodes with reflective moral agency, abstract reasoning, emotional awareness, and self-reflective choice. A human node evaluates internal impulses, chooses between constructive stewardship or destructive friction, and continuously writes updates to its own high-dimensional state record.
- **A Tree or Coral Reef as an *Environmental Balancing Unit*:** Ecosystem flora and marine structures operate as specialized stabilization hardware. A tree filters carbon, produces oxygen, stabilizes topsoil, and regulates local water cycles. A coral reef builds physical bio-shelters, buffers ocean currents, and hosts

complex marine micro-networks. Their primary systemic purpose is to maintain environmental equilibrium so that higher-level biological applications can run smoothly.

- **An Elephant as an Ecological Engine: Large animals can act as dynamic ecological sculptors.** An elephant clears dense underbrush, digs water access points during droughts, disperses seeds across vast geographic distances, and shapes entire forest and savanna biospheres. Their evolved behavioural patterns can influence nutrient cycles and ecological health.
- **A Planet as a Physical Staging Ground (Sandbox):** A planet can be treated as a macro-infrastructure node. It provides localized gravity, atmospheric conditions, chemical resources, and—on some worlds—liquid solvents such as water. Within the Cosmic Architecture analogy, a habitable planet can be pictured as a bounded 3D sandbox: a localized environment in which developing conscious entities can emerge, survive, and face constraints.
- **A Galaxy as a *Regional Network Sector*:** A galaxy is a massive structural cluster that binds billions of stellar manufacturing nodes, solar systems, and planetary sandboxes together through gravitational geometry. It functions as a regional administrative and physical sector within the larger cosmic canvas.
- **A Universe as a Self-Contained Tenant Domain:** In the model, an entire universe can be represented as a complete operational domain characterized by its physical parameters, fundamental constants, and causal rules. A hypothetical multiverse can then be pictured through the analogy of multiple tenant domains whose deeper relationships remain unknown.

By assigning a functional purpose to each scale of existence, we see that reality is not a random collection of uniform building blocks. It is an intricately layered ecosystem composed of purpose-driven, specialized participants.

2. From Isolated Entities to Interconnected Graphs

Within the model, every entity can be represented by its own distinct state vector, while its relationships are represented separately through network connections. Entities continuously interact with one another by exchanging matter, energy, and information. When two or more nodes exchange inputs and outputs, they create **relationships**.

Modern computer science and network theory represent these relationships using a mathematical structure called a **Graph**. In a graph network:

1. **Nodes** represent the individual entities (people, cells, planets, or stars).
2. **Edges** represent the active relationships, communication pathways, or physical forces linking those nodes together.

Distinguishing Similarity from Connection

To prevent confusion when mapping the geometry of reality, we must maintain a clear distinction between the **Vector State-Space** introduced in Chapter 3 and the **Graph Network** introduced here:

- **Similarity (Vector Space) asks:** *"Which entities resemble one another in state, frequency, or character?"*
- **Connection (Graph Network) asks:** *"Which entities share an active physical, social, or causal link?"*

This distinction is vital for understanding how reality operates across multiple dimensions:

Example 1: High Similarity, Zero Connection Imagine two Earth-like water planets residing in galaxies separated by ten billion light-years. Both planets possess liquid oceans, nitrogen-oxygen atmospheres, and developing biological life. In high-dimensional state-space, their coordinates sit directly adjacent to one another because their physical parameters are nearly identical (high similarity). However, because they lie outside each other's causal light cones, no photon or gravitational signal connects them. On the cosmic graph network, they share no direct edge (zero connection).

Example 2: Zero Similarity, High Connection Consider Earth and the Sun. Earth is a cold, rocky, liquid-bearing planetary sandbox hosting biological life. The Sun is a superheated nuclear fusion plasma engine. In state-space, their physical vectors are radically different (low similarity). Yet, they are tightly bound by immense gravitational metrics and continuous photon radiation. On the local solar system graph, they are connected by a massive, high-bandwidth topological edge (high connection).

Understanding both perspectives allows us to see how entities interact. Similarity determines where a node belongs in frequency and alignment, while network connections dictate how influences, energy, and karmic consequences propagate across physical and social space.

3. Overlapping Networks and the Natural Emergence of Hierarchy

Every human node simultaneously participates in dozens of overlapping graph networks:

- **Biological Networks:** Your physical organs, tissues, and cells exchange molecular signals to maintain homeostasis.
- **Ecological Networks:** Your body exchanges oxygen, carbon, water, and nutrients with local plant and microbial biospheres.
- **Social & Relational Networks:** You hold emotional, ancestral, and legal edges with family members, friends, and colleagues.
- **Economic & Civilizational Networks:** Your daily choices, labour, and transactions ripple across global trade grids and legal institutions.

Animals belong to food webs and migratory herds. Planets belong to solar systems, solar systems cluster into galactic arms, and galaxies thread along the Cosmic Web. *Reality* increasingly reveals itself not as a collection of standalone objects, but as a vast, multi-layered matrix of overlapping relationships.

As graph networks expand in size and complexity, a universal structural pattern naturally emerges: **Hierarchy**.

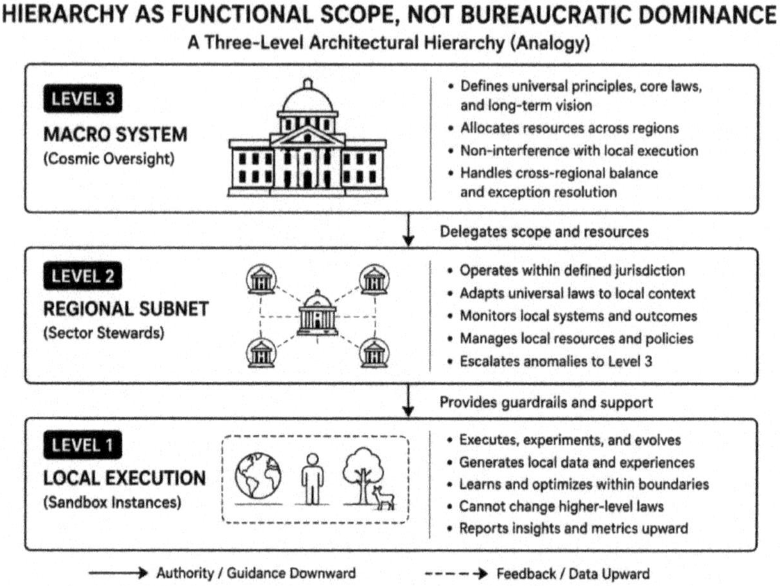

Hierarchy as Functional Scope, Not Bureaucratic Dominance
In human politics, the word "hierarchy" often carries negative connotations of power, control, or authoritarian management. But in natural systems, computer science, and cosmic architecture, hierarchy represents something entirely different: **the organization of complexity through distributed scope**.

Hierarchy emerges naturally in nature because no single level of organization can process every micro-detail simultaneously:

- A government cannot micromanage every individual sidewalk transaction in every city; it sets national policy while municipal councils handle local roads.
- A human brain does not consciously send individual electrical signals to adjust insulin levels in the pancreas or direct cardiac tissue contraction; autonomous subconscious subsystems manage biological maintenance locally.

Within the Cosmic Architecture model, hierarchy can be understood as one possible solution to information overload. It allows local execution units to operate autonomously within localized sandboxes while higher structural layers are imagined as maintaining broader stability, governance rules, and resource boundaries.

Chapter 5

Cosmic Data Pipeline, Similarity, Relationships and the Geometry of Reality

One of the most fascinating ideas from modern computer science and artificial intelligence is that **things can be "similar" to each other in hundreds of different ways at the same time.**

Think about how you describe a person. A single person is not defined by just one trait. Every human being carries dozens of observable and hidden characteristics: age, native language, hometown, job, core values, scientific curiosity, artistic taste, and emotional maturity.

None of these traits alone tells the whole story, but each one gives us a different way to measure how "close" or similar that person is to someone else:

- If you group people by **language**, you sit close to everyone who speaks your native tongue.
- If you group people by **profession**, you sit close to others in your industry.
- If you group people by **compassion and moral values**, you sit close to those who share your worldview.

In data science, when a computer maps an object across many different traits at once, it creates a **high-dimensional profile** (often called an *embedding* or *vector*). This means you don't belong to just one group—you belong to thousands of different "similarity neighbourhoods" simultaneously, depending on which trait is being measured.

1. From Data Warehouse to Vector Space: The Lifecycle Pipeline of the Soul Node

To explore how the model might represent a person as being matched to a specific family, country, or environment at birth, we can look at how modern technology companies handle vast amounts of scattered information.

In real-world software engineering, a large company like a bank or a streaming service does not run complex analysis directly on its day-to-day transaction servers—doing so would crash the system. Instead, information flows through a **three-step pipeline**:

- **Transactional Databases (Everyday Activity):** These are simple, fast systems that record immediate, raw events—like a credit card tap, a button click, or a single song play.
- **The Data Warehouse (Historical Summary):** At the end of the day, all those millions of raw events are gathered, cleaned, compressed, and organized into a central historical database.
- **The Vector Database (AI Similarity Matching):** Advanced AI converts those complex historical profiles into mathematical descriptions (embeddings) to find instant matches.

An Everyday Analogy: Think of Spotify or Netflix. Spotify records every song you listen to (Transactional Data). At night, its systems summarize your music habits (Data Warehouse). Then, its AI compares your overall taste—tempo, mood, genre, and instruments—against millions of other songs to recommend your next favourite track (Vector Database Similarity Search).

In *The Cosmic Architecture*, this enterprise pipeline becomes a helpful metaphor for how a soul's history might be gathered, summarized, and matched across lives. The model does not claim a physical computer exists in space; it borrows enterprise design to make philosophical destiny easier to understand.

ENTERPRISE PIPELINE CONCEPT | COSMIC ARCHITECTURE EQUIVALENT

Transactional Action Database | Real-Time Runtime Ledger (Daily choices & minor actions)
Karmic Debt & Impact Ledger | High-Impact Ledger (Intentional harm or deep compassion)
Master Historical Repository | Node Historical Database (Compressed multi-life records)
Data Warehouse (ETL Stage) | Master Soul Record Vault (Calculates overall net state)
Vector Database (Embeddings) | High-Dimensional State-Space (Maps Nearest Neighbours)
Nearest-Neighbour (K-NN) Query | Karmic Routing Engine (Assigns birth-place & life events)

Filtering Noise from Signal: What Gets Recorded?

A common fear is that the universe might be logging every single passing thought, minor frustration, or petty irritation into a permanent cosmic ledger. In system design, logging every piece of background noise creates unmanageable clutter and ruins system performance. Because human minds have free agency and a dynamic emotional spectrum, thousands of fleeting thoughts pass through our heads every day. The Cosmic Architecture makes a very clear distinction between **Background Runtime Noise** and **Immutable Commits**:

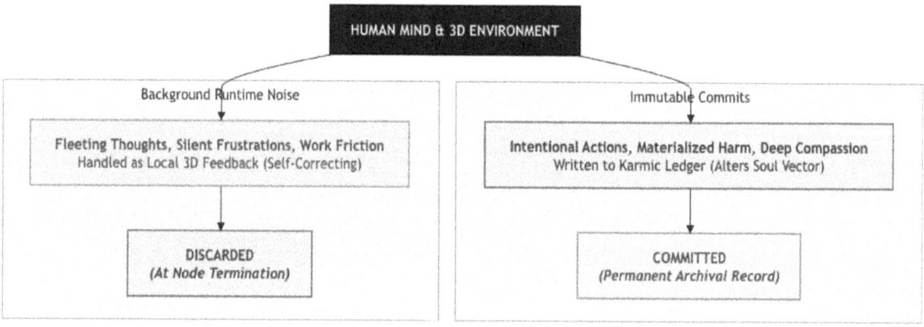

A. Unactioned Thoughts & Daily Friction (Discarded at Death)

- **Passing Anger & Silent Impulse:** Grumbling in your head when stuck in traffic, feeling temporary resentment, or having a fleeting bad thought that you choose *not* to act upon.
- **Workplace & Family Friction:** Minor arguments with colleagues, temporary quarrels with relatives, or day-to-day stress.
- **Why It Isn't Permanently Stored:** The human brain naturally processes raw emotional impulses. These *unactioned* thoughts trigger automatic local feedback loops—like brief awkwardness or temporary stress—and resolve themselves in real time. Unless an internal impulse crosses the threshold into **deliberate, materialized action or intentional harm** within this particular model it carries no permanent karmic weight and is discarded when the physical body expires.

B. The High-Impact Karmic Ledger (Permanently Archived)

- **Deliberate Actions & Materialized Intent:** Choices that result in physical, emotional, or financial harm—or conversely, acts of profound selflessness, truth, and intentional service—are written directly to the permanent record.
- **The Write-Permission Trigger: Conscious intent coupled with materialized action** acts as the write-permission trigger that permanently updates a person's master record.

Compiling the Master Soul Vector at Birth

When a physical body dies (node termination), the local 3D engine closes. Within the model, the transition can be represented as an automated summary pipeline:

1. **Extraction & Compression:** The central system queries past life trajectories, the finalized lifetime action log, and the high-impact karmic ledger.
2. **Warehouse Aggregation:** Inside the "Cosmic Data Warehouse," these scattered records are merged into a single, unified summary profile.

3. **Vector Generation (The Master Soul Vector):** The system transforms this summary profile into a multi-dimensional coordinate—a high-resolution snapshot representing the soul's overall spiritual, moral, and developmental frequency.

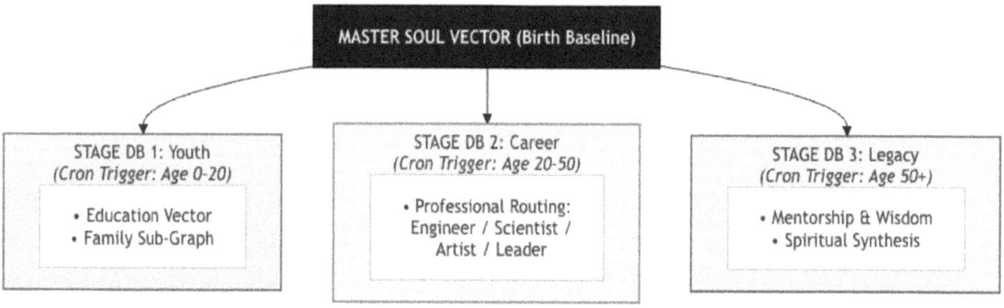

Nearest Neighbours: Determining Birth Placement and Life Trajectories

In data science, a **Nearest-Neighbour Query** finds items that share the closest matching state. Within our model, this explains how a soul finds its starting conditions at birth:

- **Initial Conditions at Birth:** Your birth circumstances—family, geography, culture, resources, and initial challenges—are matched to your Master Soul Vector. They are starting coordinates for moral growth, not mechanical punishments.
- **Dynamic Growth During Life:** Birth placement is not a rigid destiny. As you exercise free choice—acting with integrity, discipline, and compassion—your choices actively update your coordinates, opening new doors and changing your environment.

Stage-Level Databases: Life-Chapter Routing (Astrological Alignment)

Human life does not unfold as a single, static 80-year block. Ancient traditions like astrology (*Jyotisha*) noted that life shifts in distinct chapters—focusing on education in youth, career and family in adulthood, and legacy or spiritual reflection in later years.

In the Cosmic Architecture, this is modelled as **Stage-Level Vector Queries** triggered by cosmic timers which are schedulers that automatically run at different stages in lifetime.

Instead of running one giant search for your entire life, the cosmic system partitions the journey into distinct stages:

- **Stage 1: Foundation & Education (Ages 0–20):** Connects you to specific schools, mentors, and foundational learning environments.
- **Stage 2: Career & Purpose (Ages 20–50):** Triggered at adulthood, the system queries specialized professional sub-databases. If your accumulated state aligns with analytical problem-solving, it routes you toward **Engineering or Cloud Architecture**. If it aligns with theoretical discovery, it routes you toward **science**. If it carries high charisma and storytelling ability, it routes you toward **Media or Performing Arts**.
- **Stage 3: Legacy & Wisdom (Ages 50+):** Shifts focus away from purely material production toward mentorship, philosophy, and spiritual synthesis.

Real-World Tech Analogy: How AI Searches Billions of Possibilities Fast

Modern AI systems often search enormous datasets by representing each record as a **vector**– a mathematical pattern containing many characteristics at once. Instead of relying only on labels such as *engineer*, *musician* or *introvert*, the system compares the overall pattern of the record. This matters because human labels can be incomplete or wrong. A person officially classified as an accountant may also be an exceptional musician, athlete or mathematician. Vector search can represent all of these characteristics simultaneously rather than forcing the person into one rigid category.

An **HNSW graph (Hierarchical Navigable Small World graph)** is one way a vector database finds similar records quickly. Imagine 10 million profiles represented in a large multidimensional space. Rather than comparing the target with all 10 million records one by one, HNSW creates connections between mathematically similar records and uses several graph layers to navigate quickly toward the relevant region. It starts with a small number of widely connected nodes, makes large similarity jumps, then moves into denser layers where the comparisons become progressively finer. The important point is that the system is not searching fixed categories such as *Engineering → Climbing → Classical Music.* It is continually asking, **"Which connected record is mathematically closer to the target pattern?"**

For example, suppose the target profile includes software skills, astronomy, mountain climbing, quiet temperament and classical music. The search might move from a broadly technical profile to another with software and outdoor interests, then to one with astronomy and climbing, and finally to several highly similar profiles. In simplified form:

10 million records → broad similarity region → smaller neighbourhood → closest matching profiles

The Cosmic Architecture can use the same principle only as an analogy. A human or animal node would not need to be placed into perfect human-made categories. Instead, it could be represented by a much richer **multi-dimensional node signature** containing biology, temperament, abilities, experiences, relationships, intentions, actions and accumulated karmic state. DNA could be one part of that signature, but not the sole identifier; even individuals with very similar genetics can develop very different life histories and behavioural states. The cosmic match would therefore depend on the **overall evolving pattern of the node**, not a single label.

Multi-Stage Recommendation Systems

Modern recommendation systems add another useful idea. A service containing hundreds of millions of songs or videos does not perform its most expensive analysis on every item each time a recommendation is needed. It first retrieves a smaller set of broadly promising candidates, then applies more detailed contextual filtering, and finally ranks the best remaining options.

A simplified example might be:

500 million possibilities → 500 promising candidates → 20 strong matches → Top 5

The exact numbers are unimportant. The principle is **progressive narrowing.** Fast methods handle the enormous search space first; detailed analysis is reserved for the relatively small number of candidates that survive.

The Cosmic Architecture can be imagined similarly. A human node approaching an important transition would not require every possible person, city, career or event in existence to be compared at maximum detail. The architecture could first identify broadly compatible possibilities, then apply present-life context, relationships, developmental requirements and karmic state, leaving a much smaller set of plausible pathways. Free will would still determine which opportunities are accepted, rejected or never activated.

Cosmic possibility space → similarity retrieval → contextual filtering → compatible pathways → free-will choice

This makes the framework less like a system that assigns one predetermined destiny and more like one that continually exposes a changing neighbourhood of compatible possibilities.

2. Geographic Clusters and Karmic Vectors

A cosmic vector database maps more than just individual people. Lands, cities, regions, and nations are also nodes within the cosmic network, carrying their own distinct cultural and energy signatures.

Within the Cosmic Architecture model, such relocation can be interpreted as a state-space alignment:

- **Geographic/Land Layer:** Resonance with local culture, opportunities, and climate.

- **Karmic State Layer:** Specific life themes triggered by relocation—such as career growth, healing, or spiritual reflection.
- **Relational Resonance Sub-Layer (Optional):** Meeting a partner through complementary micro-vectors.
- **Lineage Sub-Graph Layer (Optional):** Bringing children into the world, creating an enduring family branch in the cosmic network.

Crucially, not every move activates every sub-layer. You might relocate to a city purely for study, recovery, or career development without initializing a family lineage. In another location, all layers align: you arrive, meet a complementary partner, and build a lasting lineage.

3. How Relationships Spawn Permanent Cosmic Sub-Graphs

When two people connect deeply, a static similarity coordinate converts into an active, dynamic graph connection. How that connection evolves depends on active vector parameters:

A. The Child as an Immutable Connection

In human society, relationships can change or end legally in divorce. However, the Cosmic Architecture distinguishes between social status and cosmic network structure:

- **Earthly/Social Layer:** Separation or divorce occurs locally as personalities diverge.
- **Permanent Cosmic Record:** The birth of a child creates a lasting biological, historical, and relational connection between two family lines. Even if parents separate on Earth, the fact of that shared lineage remains a permanent part of reality's history.

B. Lifelong Partnership Without Children

If two people spend an entire lifetime together in deep love and companionship without having children, it is not an "incomplete" relationship. The model need not interpret the absence of biological reproduction as incompleteness. Their union is a profound, self-contained partnership created for mutual elevation and shared growth.

4. Free Will, Dynamic Recalibration, and Karmic Routing

A common misconception would be to assume that vector geometry makes fate hardcoded or fixed. Within the Cosmic Architecture model, life is instead represented through an ongoing interplay between two conceptual layers:

1. **The Master Baseline Vector (Starting Line):** Cloned from past evolutionary history, setting your starting environment, genetic traits, and baseline tendencies at birth.
2. **The Active Runtime Ledger (The Free-Will Engine):** Enforces a strict Non-Obstruction Protocol. You choose, act, and intend freely at every moment.

[INITIAL BASELINE] —► Node Provisioned at Birth (Starting Line)

▼

[FREE-WILL EXECUTION] —► Real-Time Choices Made (Input / Action)

▼

[DYNAMIC RECALIBRATION] —► Incremental State Update (Output / Karma)

▼

[NEAREST-NEIGHBOR SHIFT] —► Life Events Adjust Dynamically (Environment / Fate)

Real-Time Nearest-Neighbour Recalibration

Free will sits at the **input stage**; karma and vector routing operate deterministically at the **output stage**:

- **Constructive Choices (Positive Shifts):** Consistently practicing integrity, discipline, and compassion shifts your state vector upward. Your nearest-neighbour alignment recalibrates, unlocking new opportunities, elevated relationships, and transformative environments. Karmic state influences longer-term compatibility and developmental trajectory; it does not guarantee immediate material reward or punishment. Local causality, other people's free choices, biological chance, social systems, and environmental events continue to operate.

- **Destructive Choices (Negative Shifts):** Choosing deception, harm, or cruelty instantly recalibrates your internal vector. Your nearest neighbours shift toward matching environments—such as legal friction, broken trust, or sudden life disruptions.

When physical life ends, this accumulated pattern of character, intention, and consequence is preserved as the continuing baseline for the next chapter. Nothing meaningful is wasted; every choice shapes the fabric of tomorrow.

PART I — SYSTEM ARCHITECTURE TAKEAWAYS

Primitives of the Cosmic Matrix

1. Entities Are Purpose-Driven Nodes

Reality is not composed of generic building blocks. Entities are classified by their systemic purpose. Human Nodes act as **Conscious Execution Units** with moral agency. Flora and ecosystems function as **Environmental Balancing Units**, while planets operate as bounded **3D Staging Sandboxes** that host developmental processes.

2. Inheritance and Preservation Over Decay (Git Model)

Material decay is a local property of 3D hardware. Within the model, master blueprints, species variations, and physical laws are imagined as being preserved through distributed informational structures, represented metaphorically through repository and versioning principles. Parallel universes can then be modelled as developmental branches inheriting selected baseline patterns and architectural lessons across cosmic cycles.

3. State-Space Geometry and Topological Connection

Nearness in the cosmos is governed by **state similarity** rather than physical distance alone. Relationships, family lineages, and karmic links form active graph-network edges that connect nodes across frequency, state, and developmental history. Reality is therefore organized not merely by location, but by topological and informational proximity.

4. Dynamic Karmic Routing and the Data Pipeline

Lifetime actions pass through a filtering and evaluation process analogous to an **ETL (Extract, Transform, Load) pipeline**. Raw daily noise is discarded, while intentional moral choices update the **Master Soul Vector**. Birth placement, geography, and major life stages function as multi-stage nearest-neighbour queries, continuously recalibrating in response to free-will decisions and accumulated state.

PART II - THE UNIVERSE SCIENCE DESCRIBES

Looking Back:

In **Part I**, we built our conceptual toolkit. We learned how to view entities as nodes, relationships as networks, similarity as state-space coordinates, and historical preservation through the lens of repositories and version control. We established that an analogy does not need to be literal to expose assumptions and generate sharper questions about reality.

Moving Forward:

With our analytical framework established, **Part II** grounds our inquiry in the rock-solid foundation of established physical science. We move from systems terminology to the observable cosmos—exploring four-dimensional spacetime geometry, relativistic light-cone boundaries, quantum field substrates, dark matter scaffolding, and dark energy expansion. We will trace the 13.8-billion-year history of the universe as an automated deployment pipeline, examining how fine-tuned physical parameters allow cosmic gas to forge into stars, heavy elements, and planetary sandboxes. Part II anchors our speculative model in the physical laws that govern our local universe.

Chapter 6

Spacetime, Geometry, and Causality: The Fabric of Physical Reality

Part 1: The Misunderstood Expansion and Cosmic Geometry

When examining the nature of reality, we often start with intuitive assumptions about space and time. In everyday conversation, people picture the universe as an empty, static container—a vast three-dimensional stage where physical events occur over time. When applied to cosmology, this leads to the popular misconception that the Big Bang was a gigantic firecracker explosion occurring in the middle of pre-existing dark void, scattering matter outward into empty space.

Modern physics presents a completely different and far more precise picture: **the Big Bang was not an explosion *in* space, but the rapid expansion *of* space itself.**

To understand physical reality, we must abandon the idea that space is empty non-existence. In classical mechanics, Sir Isaac Newton treated space as a rigid, absolute background and described gravity as an invisible force pulling masses toward one another across distance. Centuries later, Albert Einstein unified three spatial dimensions (length, width, height) with time into a single, dynamic four-dimensional continuum known as **Spacetime**.

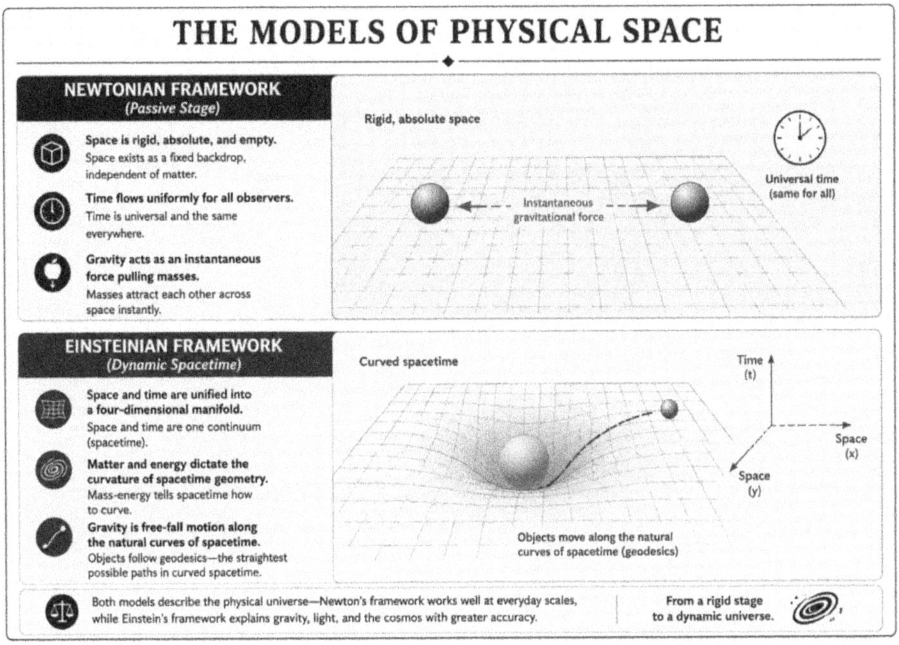

In General Relativity, gravity is not an invisible rope pulling objects across a void. Matter and energy alter the curvature of spacetime geometry, and that geometry dictates how matter and light travel. When a planet orbits a star or light bends near a massive galaxy, the objects are simply following the natural, unforced geometry of curved spacetime. In General Relativity, spacetime is modelled as a dynamic geometric structure whose curvature can be measured through its effects on matter and light.

When scientists discuss cosmological inflation, they are describing a proposed period of extraordinarily rapid expansion in the very early universe. Distant points could separate faster than light because the metric of space

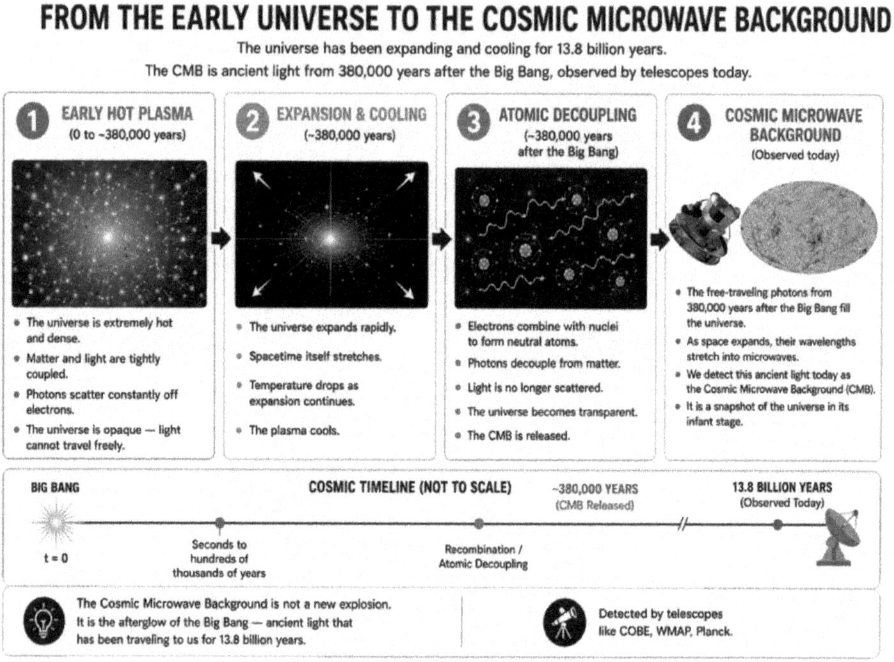

itself was expanding; this does not mean matter or information locally travelled through space faster than the relativistic speed limit.

Inflation was proposed in part to address two major puzzles. First, it can help explain why widely separated regions of the observable universe have nearly the same temperature: regions that are now far apart could once have been in causal contact before rapid expansion separated them. Second, it explains why space looks smooth and flat today, much like how blowing up a wrinkled balloon smooths out its surface.

During this rapid growth spurt, tiny, microscopic ripples in subatomic energy were stretched across vast cosmic distances. As the expansion slowed down, these stretched ripples created slightly denser patches of matter. Over time, gravity pulled gas and dust into these patches, using them as structural seeds to build the very first stars, galaxies, and cosmic networks we see today.

The Early Universe: When Light Could Not Travel Freely

For roughly the first **380,000 years**, the universe was extremely hot. It was not yet filled with ordinary atoms like those that make up the world around us today.

To understand why, we first need a very simple picture of an **atom**.

Everything around us—air, water, rocks, plants, and our own bodies—is made from atoms. An atom is an extremely tiny building block of ordinary matter. At its centre is a small core called the **nucleus**. The nucleus contains **protons**, which carry a positive electrical charge. Around the nucleus are much lighter particles called **electrons**, which carry a negative electrical charge.

You can think of positive and negative electrical charges as two opposite properties that attract each other. Under suitable conditions, negatively charged electrons can therefore remain associated with positively charged atomic nuclei, producing stable atoms.

But the early universe was **far too hot and energetic for this arrangement to remain stable**.

Here, "hot" means more than simply being hot like an oven or a flame. At extremely high temperatures, particles carry enormous amounts of energy and move and interact intensely. Electrons could not remain stably attached to atomic nuclei. Instead of ordinary neutral atoms, the universe was filled with a hot mixture of free electrically charged particles.

Physicists call this state of matter **plasma**.

Plasma may sound exotic, but it is not merely theoretical. The Sun and other stars contain enormous amounts of plasma. The early universe was also dominated by this charged state of matter.

This created an important problem for **light**.

Light was present, but it could not travel freely across great distances. The many free electrons repeatedly interacted with and scattered the light. A useful everyday comparison is trying to shine a torch through extremely dense fog. The torch is producing light, but droplets in the fog scatter it in many directions, preventing you from seeing clearly through it.

The early universe was obviously not literally filled with water fog. The comparison simply helps us picture why light could exist without being able to travel freely over enormous distances.

As the universe **expanded**, however, it also **cooled**. Eventually it became cool enough for electrons to remain associated with atomic nuclei, producing stable, electrically neutral atoms—mostly hydrogen.

This transition is known as **recombination**.

The name can be slightly confusing because it sounds as though electrons and nuclei must previously have been combined and were now being combined *again*. In cosmology, however, *recombination* is simply the conventional name for this early period when stable neutral atoms formed.

Once most free electrons were bound into atoms, light encountered far fewer charged particles capable of continually scattering it. The cosmic "fog" effectively cleared, and light could travel much more freely through space.

Remarkably, we can still detect that ancient light today.

During the billions of years since it was released, the universe has continued to expand. As space expanded, the light travelling through it was stretched to longer **wavelengths**. A wavelength is simply the distance between repeating peaks of a wave. Different wavelengths of light correspond to different parts of what physicists call the **electromagnetic spectrum**.

By the time this ancient light reaches our instruments today, it has been stretched far beyond visible light into the **microwave** region of that spectrum.

Astronomers therefore call it the **Cosmic Microwave Background (CMB).**

The CMB is an extremely faint glow arriving from every direction in the sky. It provides scientists with something extraordinary: a kind of **ancient snapshot of the universe from when it was only about 380,000 years old.**

In simple terms, the sequence is:

Extremely hot universe → free charged particles → light repeatedly scattered → universe expands and cools → stable atoms form → cosmic fog clears → light travels freely → ancient light eventually becomes the CMB we observe today.

Part 2: Cause and Effect, the Speed of Light, and Spacetime

Before going further, we need one simple rule: **a cause must be able to reach whatever it affects.**

Imagine the Sun suddenly changed. Earth could not know about that change instantly because information from the Sun takes about eight minutes to reach us. Nothing carrying ordinary information can simply jump across space instantaneously.

This brings us to the **speed of light.**

The speed of light is not important only because light happens to travel at that speed. According to Einstein's theory of relativity, it represents the maximum speed at which information and cause-and-effect influences can locally travel through space.

In simple terms:

No information from here can affect somewhere else faster than this cosmic speed limit allows.

💡 Core Physics Insight — Space and Time Belong Together

We normally think about **space** and **time** separately.

If you meet someone for coffee, for example, you need two pieces of information:

Where? — the café.

When? — perhaps 10:00 a.m.

Neither piece of information is enough by itself. Together, they describe an **event**.

Einstein's relativity tells us that the universe works in a similar way. Space and time are interconnected into what physicists call **spacetime**.

Different observers moving at different speeds can disagree about how much distance or time separates two events. Relativity nevertheless provides a combined measure of space and time—the **spacetime interval**—on which all inertial observers agree.

The speed of light provides the fundamental relationship connecting these spatial and temporal measurements.

For a general reader, however, the most important idea is much simpler:

Spacetime determines which events can possibly influence which other events.

Physicists represent this using something called a **light cone**.

Imagine standing at one point in space and switching on a powerful light.

As time passes, that light spreads farther outward. If we draw this expanding region together with the passage of time, it forms a cone-shaped diagram.

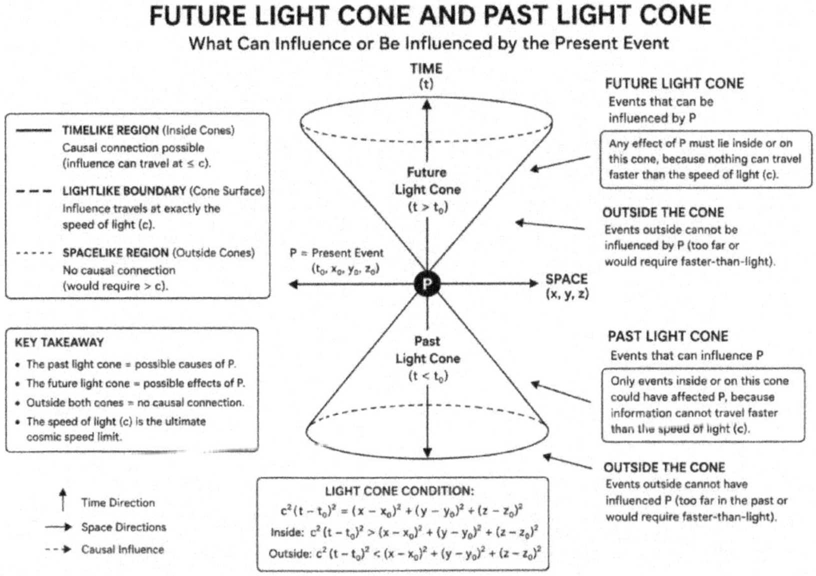

1. **Future Light Cone:** Everything you could potentially influence from this moment, because there is enough time for light—or something slower than light—to reach it.
2. **Past Light Cone:** Everything that could potentially have influenced you, because light or information from those events had enough time to reach your present location.
3. **Outside the Light Cone:** Events that are simply too far away for information to travel between them in the available time. Physicists call these events **spacelike separated**.

A simple way to picture it is:

Past events that could reach you → YOU NOW → Future events you could eventually reach

Anything outside those boundaries cannot exchange ordinary cause-and-effect information with you at that moment.

Why Does the Speed Limit Matter?

This speed limit is important because it helps preserve the basic order of **cause and effect**. Normally, a cause happens first and its effect follows—you push a glass, and then it falls. However, Einstein's relativity shows that space, time, and motion are interconnected. If controllable information could travel faster than light, there are theoretical situations in which a message could be sent to a distant location, and a reply could effectively arrive **before the original message was sent**. This could create a **causal loop**, where an effect appears before its own cause, leading to the familiar paradoxes associated with time travel. This is one reason the speed of light is more than simply the speed at which light travels: it also forms a fundamental boundary that protects the normal ordering of cause and effect in spacetime.

When considering higher dimensions or additional coordinate frameworks, standard physics requires that any valid mathematical model respect these local cause-and-effect boundaries. A higher dimension is not simply "another room" appended to physical space; in mathematics and physics, a **dimension** represents an independent coordinate parameter needed to specify a position, state, or degree of freedom within a system.

Just as a point on a line can be located using one coordinate **(x)**, a point on a flat surface requires two **(x, y)**, and a point in ordinary three-dimensional space requires three **(x, y, z)**. More complex physical systems may require many more values to describe their complete state. Physicists represent these possibilities using mathematical spaces with many dimensions, often called **high-dimensional state spaces**.

Summary of Foundational Physics Concepts

To ensure consistency across all analytical discussions, the core physical parameters of space, geometry, and causality are summarized below:

PHYSICAL PRIMITIVE	STANDARD DEFINITION & FUNCTION	SYSTEMIC RULE IN REALITY
Spacetime	The 4D continuous manifold uniting 3 spatial dimensions and 1 time dimension.	Acts as a dynamic geometric fabric curved by mass and energy, rather than an empty void.
Cosmological Inflation	Rapid metric expansion of space in the early universe.	Explains cosmic flatness, horizon uniformity, and the density seeds of large-scale structure.
Gravitational Curvature	Motion along geodesics in curved spacetime geometry.	Replaces "action-at-a-distance" forces with natural geometric paths.
Invariant Speed	The maximum speed of information transfer in spacetime.	Establishes the structural boundary of causality and connects space with time.
Light Cones	Geometric boundaries defining past and future causal horizons.	Prevents spacelike paradoxes by restricting which events can physically influence others.

PHYSICAL PRIMITIVE	STANDARD DEFINITION & FUNCTION	SYSTEMIC RULE IN REALITY
Higher Dimensions / State-Spaces	Additional independent coordinate axes needed to define complex system states.	Provides mathematical parameters for complex state-spaces without violating local 4D causality.

Part 2: Time Travel, State Rollbacks, and the Append-Only Cosmos

Time Travel, State Rollbacks, and the Append-Only Cosmos

Few concepts in theoretical physics capture the human imagination like time travel. Science fiction frequently depicts the past as a physical room sitting down an invisible hallway in space—a destination one could visit simply by building a fast enough ship or finding a clever shortcut through spacetime.

When we evaluate time travel through both relativistic physics and the Cosmic Architecture, a clear distinction emerges between moving *forward* into the future and attempting to travel *backward* into the past.

A. Forward Time Dilation: Variable Execution Speeds

Moving forward through time at different rates is a proven physical effect described by Einstein's theories of relativity. Time is not a universal metronome ticking at exactly the same rate for every observer.

Gravitational and Velocity Time Dilation: Time can pass at different rates for observers depending on their motion and the gravitational fields they experience. An observer travelling at a speed close to the speed of light, or spending time in a sufficiently strong gravitational field, may experience **less elapsed time** than another observer who remains farther away or moves differently.

Importantly, the travelling observer does not feel their own time slowing down. Their heartbeat, thoughts, watch, and everyday experiences continue normally. The difference becomes apparent when the two observers later compare how much time has passed for each of them.

In system-design terms, we can picture forward time dilation as **variable clock-rate processing**. Imagine, for example, that an astronaut spends one year according to their own clock in circumstances where ten years pass for distant observers on Earth. The astronaut has not broken the universe's causal order or jumped outside time. Their path through spacetime simply accumulated less elapsed time. If they later return to Earth, they would arrive in an Earth that had aged much more than they had—but they could not use the same process to travel backward into Earth's past.

B. Backward Time Travel: The Append-Only Ledger

While relativity's field equations allow theoretical geometric shortcuts like Closed Time like Curves (CTCs) or wormholes, backward time travel introduces severe logical contradictions—most notably **causal paradoxes** (such as preventing one's own past existence). Physicists like Stephen Hawking proposed the *Chronology Protection Conjecture*, suggesting that quantum feedback loops naturally destroy time loops before they can form.

The *Cosmic Architecture* provides a structural reason why physical travel to the past is impossible: The universe is an append-only, state-updating runtime environment.

The physical year 1920 does not exist as a physical location in deep space waiting to be visited. The physical atoms that composed 1920 Earth were repurposed through thermodynamics and stellar/geological cycles to build 1950 Earth, 2000 Earth, and 2026 Earth. As established in Chapter 19 (*Cosmic Memory*), **matter itself is the storage medium**. The present physical state of the universe *is* the unalterable record of its history.

C. The Fallacy of System Rollbacks

For an observer in 2026 to physically travel to 1920 Earth, the cosmos would have to execute a **Full System State Rollback**:

1. The active 2026 runtime engine would have to freeze.
2. Every atom, planet, flora, fauna, and human in the universe would have to be un-spun and reallocated back to its exact 1920 state vector coordinates.

This exposes a basic misunderstanding about time. Science fiction often treats past eras as hidden coordinates that advanced spacecraft could locate and visit. No sensor or engine can target a previous state of the universe. Even if the cosmos somehow rolled back its entire system to 1920, the active runtime would be overwritten. The 2026 observer, their future knowledge, and their spacecraft would disappear from the record, leaving no one to know that the rollback had occurred.

D. How the Cosmos Interacts with History: Queries and Branching

Within the *Cosmic Architecture*, rolling back an active universe would be an inefficient and disruptive way to revisit history. It would potentially erase or overwrite the accumulated choices, relationships, karmic states, and developmental progress of countless conscious nodes.

Modern relativity also gives us a useful way to think about time. In one interpretation, often called the **block-universe view**, spacetime can be represented as a four-dimensional structure in which events occupy different locations not only in space but also in time. This does not demonstrate that the past exists as a cosmic database that can literally be accessed. However, it provides a useful conceptual foundation for asking whether history might be preserved without requiring physical travel back into it.

Within the speculative *Cosmic Architecture*, history can therefore be imagined as being handled through three different mechanisms:

- **Read-Only Historical Audits:** Light travels at a finite speed **(c)**, which means astronomy already allows us to observe the past.

When we look at the Sun, for example, we see it as it was roughly eight minutes earlier because its light takes time to reach Earth. When astronomers observe a galaxy millions of light-years away, they see that galaxy millions of years in its past. The *Cosmic Architecture* extends this established principle speculatively: if deeper informational traces of events are preserved within reality, higher governance tiers—or elevated states of consciousness such as *Kevala Jnana* within the model—might access historical information without physically travelling backward in time. In systems language, this resembles a **read-only telemetry query** rather than a rollback of the active environment.

- **Systemic Branching (Parallel Sandboxes):** If a higher governance tier needed to explore a historical "What If?" scenario, the model does not require the primary universe to be rewound. Following the version-control analogy introduced in Chapter 2, a separate **branch or sandbox** could instead be initialized from an earlier state and allowed to develop independently. This is a speculative architectural mechanism, not an established feature of cosmology, but it avoids altering the history of the original universe.
- **Targeted Node Migration:** The same principle can be applied to conscious development. If, within the model, a persistent conscious node carries unresolved karmic or developmental requirements, history itself need not be reversed. The node could instead be re-embodied within another compatible environment whose conditions provide opportunities for continued development. Again, this is a metaphysical component of the *Cosmic Architecture*, rather than a conclusion of modern physics.

The resulting architecture treats **history as preserved state rather than editable state**. The past records what has occurred; the present is where new choices and interactions take place; and the future remains the direction in which new states can emerge.

In the software analogy, history resembles an **immutable sequence of committed states**. We may inspect an earlier commit, and we may even create a new branch from it, but we do not rewrite the production history that led to the current state.

Likewise, the *Cosmic Architecture* does not require conscious beings to travel backward along their own histories. Relativity, causality, and the thermodynamic arrow of time all place powerful constraints on ordinary physical processes, while no experimentally demonstrated mechanism currently allows humans to travel into and alter their own past. The model therefore treats backward physical time travel as unnecessary: **history can be read, alternative possibilities can be explored through branches, and conscious development can continue through new states without rewriting what has already occurred.**

How Chapter 6 Functions as the Physical Anchor for the Book

Chapter 6 establishes the rigorous baseline of **known physical law** that serves as the foundation for the rest of this volume:

1. **Establishes Hard Causal Boundaries:** By defining light cones and the invariant speed, this chapter sets the standard that any system or interaction discussed in later chapters must account for cause-and-effect structure.
2. **Clarifies Geometric Reality:** By establishing that space is a dynamic metric rather than a passive box, it grounds all subsequent discussions of network topologies, vector spaces, and state-spaces in formal geometric principles.
3. **Defines Higher Dimensions Rigorously:** By defining dimensions as independent mathematical coordinates rather than vague "magical realms," it provides a clean, scientific vocabulary for analysing multi-layered models across future chapters.

Chapter 7

Quantum Reality Without Mysticism – The Invisible Cosmic Network

Note: *Standard quantum mechanics and general relativity establish the physical laws of our 3D observable universe. In this chapter, principles like entanglement and bulk fields are explored as speculative systems analogies for higher-dimensional connectivity, rather than proven empirical messaging channels.*

Quantum physics is often treated as either a dry, mathematical nightmare reserved for laboratory scientists, or a magical, mystical phenomenon where "anything goes." Both extremes miss the bigger picture.

One way to read quantum mechanics within this book's framework is as a possible information-bearing layer of physical reality. This does not mean quantum theory proves a cosmic network or mystical architecture. It means quantum physics gives us unusually rich language for thinking about fields, probability, measurement and non-classical correlation.

To explore these questions, we do not need to memorize complex physics equations. We first need a simple understanding of how quantum physics describes information, probability, fields, measurement, and relationships across physical distance.1. What Is Quantum Physics?

Before discussing quantum fields, entanglement, or hidden connections, we first need to understand what **quantum physics** actually means.

The world we experience every day seems predictable. A football has a location, a car follows a path, and if you drop a glass, gravity pulls it toward the floor. This everyday world is described extremely well by what physicists broadly call **classical physics**.

But when scientists began studying nature at extremely small scales—atoms, electrons, photons and other subatomic particles—they discovered that the familiar rules of everyday objects were no longer enough.

An **atom** is one of the basic building blocks of ordinary matter. At its centre is a tiny nucleus containing **protons** and **neutrons**, while **electrons** are described by quantum states associated with the atom. A **photon** is a quantum of electromagnetic radiation—the kind of particle-like excitation associated with light.

Quantum physics is the branch of physics developed to describe how nature behaves at these very small scales.

What makes it unusual is that quantum physics often describes **possible outcomes and their probabilities** rather than simply assigning every object an ordinary definite path in advance. Quantum systems can also display phenomena such as **superposition** and **entanglement**, which have no simple equivalent in our everyday experience.

A useful way to understand the difference is:

Classical world:
"The object is here and is following this path."

Quantum description:
"The system is in a state that determines the probabilities of the different results we could obtain when we measure it."

This does not mean that *anything can happen*. Quantum mechanics follows extremely precise mathematical rules and is one of the most accurately tested theories in science. The strange part is that those rules describe reality very differently from the way human intuition expects the world to behave.

1. Why the Cosmos Needs Qubits: The Engine of Possibility

In everyday human computing, the smallest unit of information is a bit. A bit is rigid; it has only one of two values: 0 or 1. You can imagine it like a light switch that is either permanently OFF or permanently ON. If the universe ran on classical bits, reality would be completely predetermined—like a movie playing on a DVD. Every event, every human choice, and every leaf falling would be hardcoded in advance, leaving no room for free will, moral agency, or dynamic change.

But the Cosmic Architecture is not a pre-recorded movie; it is a live, dynamic system. To process the unimaginable complexity of billions of conscious minds making choices simultaneously, the universe requires a vastly superior processing unit: the **qubit** (quantum bit).

Unlike a rigid bit, a qubit can exist in a "superposition"—meaning it can hold multiple possibilities at the exact same time before it is measured or acted upon.

The Everyday Analogy: Imagine you are using a GPS app to drive home. A classical bit is like a printed map with a single route drawn in pen; your path is fixed. A qubit is like the live GPS algorithm constantly calculating ten different possible routes simultaneously, adjusting in real time based on traffic. It holds all those possibilities open *until you actually turn the steering wheel.*

The Cosmic Application: If the cosmic architecture used classical bits, the universe would be a completely predetermined machine—a rigid clockwork where every action you take was pre-calculated at the Big Bang. But the purpose of the Earth sandbox is to test **moral agency**. For free will to genuinely exist, the universe requires a processing unit capable of holding *potential.*

In the Cosmic Architecture, qubits act as this dynamic rendering engine. When a human node approaches a moral crossroads—for example, deciding whether to act with integrity or deceit—the cosmic system does not pre-select the outcome. Instead, it holds all possible choices, and their resulting karmic trajectories, open in a suspended quantum state.

The human mind is the trigger. The exact moment you make a conscious choice, your free will acts as the "measurement." That measurement instantly collapses the infinite field of possibilities, locking your chosen action into the permanent 3D physical ledger and updating your Soul State Vector. Qubits are the essential architecture that allows the cosmos to remain entirely fluid and responsive until the very second you decide who you are going to be.

2. Quantum Fields: The Invisible Cosmic Network

In everyday life, we tend to think of the universe as a collection of hard, isolated objects—like billiard balls on a table. If a billiard ball wants to affect another one, it must roll across the table and physically crash into it. If the cosmos operated this way, information could only travel through slow, physical collisions.

But if reality is a highly interconnected architecture where your karmic choices, soul state, and nearest-neighbour alignments update continuously, the universe needs a way to transmit that information instantly across vast distances without laying physical cables.

Modern physics reveals that the universe uses **quantum fields**. The universe is not a massive empty box filled with isolated marbles. It is composed of continuous, invisible fields that span across all of space. What we call "particles" (like electrons or photons) are actually just localized ripples on these underlying fields.

The Everyday Analogy: Think of a quantum field like the water in an enormous swimming pool, and physical particles as the waves. If you splash the water at one end of the pool, a floating cork at the far end will eventually bob up and down. You didn't throw a physical object at the cork; you manipulated the shared medium connecting you both.

The Cosmic Application: In human technology, if a bank branch in London updates your account balance, the server in Tokyo knows instantly because they are connected via a secure, underlying network layer. They do not need to physically mail a ledger across the ocean.

In the Cosmic Architecture, quantum fields serve this exact function: they are the invisible, omnipresent synchronization network that touches every atom, planet, and conscious node simultaneously.

When you make a significant moral choice and your internal State Vector is updated, that new data does not need to physically travel through empty space. Instead, your updated state creates a subtle frequency shift—a ripple—directly within these foundational quantum fields. Because the entire cosmic database rests on this shared substrate, the system can instantly read your new karmic frequency. This explains the mechanics behind "Nearest-Neighbour" alignment: the quantum field seamlessly shifts your trajectory, subtly drawing new life events, challenges, and relationships into your path without a single physical wire ever being crossed.

By explaining it this way, we give the reader a tangible mental model of *how* the universe physically calculates and responds to their daily life.

4. Entanglement: Why Physical Distance May Not Tell the Whole Story

Why discuss **quantum entanglement** in the Cosmic Architecture? Because it provides a real example from physics showing that two systems can share a relationship that cannot be understood simply by imagining them as two independent objects with ordinary pre-existing properties.

When two particles become entangled, they are described by a **shared quantum state**. If scientists later measure them, even when they are very far apart, their results can display remarkably strong correlations. Experiments have shown that these correlations cannot be explained by ordinary classical models in which each particle simply carries its own hidden set of local instructions.

This does **not** mean that Particle A sends an instantaneous message to Particle B. Entanglement cannot be used for ordinary faster-than-light communication. Instead, the important point for this book is that

physical separation does not necessarily mean complete independence.

When interpreted through the Cosmic Architecture as a deliberately speculative metaphor:

- **In ordinary 3D space:** Particle A and Particle B may appear extremely far apart.
- **At the quantum level:** They can still belong to a single shared quantum description.
- **In the Cosmic Architecture analogy:** This suggests a way of imagining apparently distant nodes as deeply related within a larger state-space or network structure.

Imagine two puppets hanging through separate holes in a ceiling. From inside the room, the puppets appear separate and may even be far apart. But if we could see above the ceiling, we might discover that their strings belong to a larger structure that was invisible from below.

The analogy does **not** mean that quantum particles are literally connected by hidden strings, nor does physics show that they are nearest neighbours in a higher spatial dimension. The purpose is to illustrate a broader idea: **what appears separate when viewed from one level of reality may be more deeply related when described at another level.**

This is why entanglement is useful to the Cosmic Architecture. Later in the book, humans, planets, universes, and other entities may appear enormously separated in ordinary physical space while still being represented as neighbours or related nodes in a much larger **state-space**. Quantum entanglement does not prove that such a cosmic network exists, but it provides a powerful real-world reminder that **distance and relationship are not always as simple as everyday intuition suggests**.

5. Science, Spirituality, and the Boundaries of Thought

Quantum physics is one of the most successful scientific theories ever developed, but it is also one of the most frequently misunderstood. Terms such as **"energy," "vibration," "quantum," and "entanglement"** are sometimes borrowed to support spiritual claims that the physics itself does not actually establish. For example, quantum mechanics does not show that positive thoughts can directly reshape external reality, that consciousness can communicate instantly across the universe, or that human beings can control matter simply through intention.

At the opposite extreme, it would also be unnecessary to conclude that because science can describe a physical mechanism, all deeper philosophical questions have therefore been answered. Knowing **how** something operates and asking **why such a system exists at all** are different kinds of questions.

Science is exceptionally powerful at investigating the **mechanism**. It can describe how particles interact, how fields behave, how quantum probabilities predict measurement outcomes, and how physical laws operate with extraordinary mathematical precision.

Philosophy and spirituality can ask a different set of questions: **Why do these laws exist? Why is the universe mathematically ordered? Why does reality permit complexity, life, and consciousness? Is this structure simply a fundamental fact of existence, or could there be some deeper source, principle, or architecture behind it?** These questions do not automatically require a designer, but neither are they answered simply by describing the physical mechanism.

This distinction is important throughout *The Cosmic Architecture*. The book does not argue that quantum physics proves God, souls, karma, higher dimensions, or a hidden cosmic network. Nor does it claim that unexplained physics should automatically be treated as evidence for spirituality.

Instead, quantum physics provides something more modest but still fascinating: **a scientifically established example of reality behaving in ways that are very different from ordinary human intuition.** At the quantum scale, probabilities, correlations, and interactions cannot always be understood using the simple picture of separate, solid objects behaving like tiny billiard balls.

The Cosmic Architecture uses these discoveries as **conceptual inspiration rather than scientific proof**. If nature already teaches us that everyday intuition is not sufficient to understand its deepest physical layers, then it is reasonable to remain intellectually open to the possibility that reality may contain structures or relationships that our present understanding has not yet uncovered.

That possibility is where science ends, philosophical exploration begins, and **The Cosmic Architecture asks the reader to think—not to believe without evidence.**

Chapter 8

Dark Matter, Dark Energy, and the Cosmic Web – The Invisible Scaffolding

When we look up at the night sky, we see twinkling stars, glowing nebulae, and distant galaxies. It is natural to assume that this visible light represents the entirety of the cosmos. However, modern astronomy reveals a shocking reality: everything we can see, touch, and measure—from planets to human bodies—makes up only about 5% of the total universe.

The remaining 95% consists of two mysterious components that scientists call **Dark Matter** and **Dark Energy**. The word "dark" here does not imply something evil or supernatural; it simply means "unseen" or "unknown."

Within the Cosmic Architecture, dark matter and dark energy can be treated as powerful metaphors for hidden structure and large-scale system behaviour. Scientifically, they remain active areas of research; the model does not claim to identify their ultimate nature but uses their observed effects to ask what unseen scaffolding may mean in a broader philosophical sense.

1. Dark Matter: The Invisible Scaffolding

Galaxies across space spin at incredible speeds. According to standard laws of gravity, the visible mass of stars and gas in a galaxy does not provide enough gravity to hold it together. By all physical calculations, galaxies should fly apart like spinning merry-go-rounds.

Yet, they remain intact. Astronomers realized that an invisible form of mass surrounds and permeates every galaxy, providing the extra gravitational pull needed to hold them together. They named this unseen mass **Dark Matter**.

In our *cosmic architecture* model:

- **Dark Matter acts as the Invisible Grid / Scaffolding:** Just as an architect builds temporary steel framing before pouring concrete, the universe established invisible dark matter filaments early in its history.

- **Regular Matter Follows the Grid:** Gas, dust, and stars were drawn along these dark matter threads by gravity, pooling at major intersections to build galaxies and solar systems.

This helps explain why galaxies are not scattered randomly in space like spilled sand. They are organised along an intricate, interconnected pattern called the **Cosmic Web**. In this book, that web becomes a suggestive image of layout and structure, not proof that the universe was consciously engineered in the same way a human architect designs a building.

2. Dark Energy: System Expansion and Resource Allocation

While dark matter contributes gravitationally to the formation and binding of cosmic structure, dark energy is the name given to whatever is responsible for the observed acceleration of cosmic expansion.

Astronomical observations indicate that the expansion of the universe is accelerating. In current cosmological models, the unknown component or effect associated with this acceleration is called dark energy.

In a system design framework, dark energy reflects the dynamic balance between local stability and overall network expansion:

- **Local Clustering (Dark Matter):** Ensures that solar systems and galaxies remain stable enough for planets, chemistry, and biological life to develop.
- **Macro Expansion (Dark Energy):** In the systems analogy, accelerated expansion can be pictured as increasing the separation between sufficiently distant cosmic regions. This is a metaphorical interpretation, not a scientific claim that dark energy exists to prevent universal collapse or to maintain 'safe' distances.

3. Mechanisms as Part of Design

A common tension in conversations about cosmology is the relationship between mechanism and meaning. Scientific explanations show how structures arise from gravity, density fluctuations and physical law. Philosophical or theological interpretation asks a different question: whether the existence of such law-governed order points beyond mechanism alone. Knowing how an engine works does not settle every question about why the engine exists, what larger purpose it serves, or whether its order is ultimately brute fact, necessity, chance, or design.

- **Science explains the mechanism:** It maps the growth of filaments, gravity, and density fluctuations.
- **Cosmic Architecture explores possible meaning:** It asks whether rules, laws and initial conditions capable of organising matter into conscious beings may point to a deeper order beyond mechanism alone.

The Cosmic Web therefore demonstrates something both scientific and philosophically fertile: simple rules operating over immense time can generate astonishing order. Whether that order is interpreted as purposeless emergence, deep mathematical necessity, or evidence of a larger mind remains the interpretive question this book explores.

Chapter 9

Stars, Elements, and Planetary Systems – The Automated Cosmic Deployment Pipeline

A Note on Pipeline Terminology: *As established in Chapter 2, repository names, branches, version tags, pipelines, and deployments are human engineering notation rather than claims that the cosmos literally operates Git or software infrastructure. They provide a compact way of representing how a hypothetical distributed cosmic architecture might preserve configurations, coordinate dependencies, and progressively realize complex structures across cosmic time.*

The early universe was extraordinarily simple. Shortly after the initial cosmic expansion, reality contained virtually no complex chemistry—only simple, light gases, primarily hydrogen and helium. There was no Earth, no oxygen to breathe, no carbon for organic life, and no solid ground.

How did a cosmos filled with simple gas evolve into a world of oceans, trees, animals, and human minds?

In basic religious views, divine creation is often pictured as a creator handcrafting individual rocks, trees, and planets one by one. Conversely, strict materialist interpretations may assume that because stellar evolution follows physical laws, no deeper purpose or architecture is required. A more careful systems reading avoids both extremes. **Physical laws can generate extraordinary complexity automatically over deep time, while philosophy can still ask whether such lawful emergence reflects deeper order,** purpose, or design. Understanding how the process works does not, by itself, determine why such a law-governed system exists or whether it possesses any ultimate purpose.

1. Understanding the Cosmic CI/CD Pipeline

In modern software engineering, complex systems are never built or launched all at once in a single massive leap. Instead, they are created, tested, and updated through an automated workflow known as a **Continuous Integration and Continuous Deployment (CI/CD) Pipeline.**

Think of a CI/CD pipeline like a modern automobile assembly line. Rather than attempting to assemble an entire car instantly out of thin air, the factory operates in a strict, automated sequence:

1. Setting up the steel chassis,
2. Installing the engine and mechanical transmission,
3. Testing the electrical wiring and computer sensors,
4. Applying the protective paint coating, and
5. Rolling the finished, road-ready car off the assembly line.

In software development, a CI/CD pipeline continuously pulls code from a **Source Repository** (as introduced in Chapter 2), verifies that new updates compile smoothly, and deploys the application step-by-step to production servers without crashing the running platform for example a website.

A standard software deployment pipeline operates through distinct stages:

- **Stage 1: Infrastructure Provisioning:** Setting up the basic physical servers, memory limits, and network firewalls.
- **Stage 2: Environment Setup:** Preparing the operating system, background databases, and operational rules.
- **Stage 3: Application Deployment:** Releasing the actual software programs (microservices) onto the hardware.
- **Stage 4: Automated Maintenance & Cron Jobs:** Running scheduled updates, health checks, and garbage collection to clear out old technical debt.

When we examine the 13.8-billion-year history of our observable universe, it can be helpful to compare cosmic development to a staged pipeline: early physical conditions enable stars, stars create heavy elements, heavy elements allow rocky planets, and some planets become environments where chemistry can become biology. The analogy is powerful, but it remains an analogy; science describes these stages through physical processes rather than literal repository pulls or deployment scripts.

2. Bridging Chapter 2 & Chapter 9: Repositories and Release Tags

To see how the cosmic pipeline operates within speculative Cosmic Architecture model, we must connect it to the version control concepts established in Chapter 2:

- **The Cosmic Repositories:** Store the unalterable master blueprints for physical laws, fine-tuned constants, chemical interaction tables, and species templates.
- **Semantic Versioning (SemVer):** Every component deployed by the pipeline carries a release tag (e.g., **MAJOR.MINOR. PATCH).** A Major tag update (v1.0 to v2.0) represents a fundamental structural overhaul, while Minor (v2.1) and Patch (v2.1.4) tags represent feature additions and local environmental optimizations.
- **Multiverse Branching:** The CI/CD pipeline executes across specific universe branches. A pipeline running in a young, developing universe might deploy early species tags, while a pipeline running in an advanced multiverse sector deploys higher-tier release tags.

3. Dark Matter & Dark Energy: The Invisible Scaffolding and Dynamic Buffer Engine

In Stage 2 of the pipeline, the universe deploys two fundamental mechanisms that form the backend infrastructure of our physical reality: **Dark Matter** and **Dark Energy**.

- **Dark Matter (The Invisible Layout Grid):** In standard cosmology, dark matter is inferred from gravitational effects and plays an important role in the growth of large-scale structure. Its gravity helped matter collect along the filamentary Cosmic Web from which galaxies and clusters developed. Within this book's systems analogy, that role can be pictured as an invisible layout grid or structural scaffold. The analogy should not be taken to mean that dark matter is literally engineered infrastructure.

- **Dark Energy (Global Expansion & System Bounds):** On large scales, dark energy is associated with accelerated cosmic expansion. Within the architecture analogy, that expansion can be pictured as increasing separation between distant sectors. Physics does not establish that dark energy exists to prevent collapse, enforce isolation, or create multiverse boundaries; those are interpretive extensions of the model.

Together, dark matter and dark energy play very different roles in current cosmological models: dark matter contributes gravitationally to structure formation, while dark energy is associated with accelerated expansion. The book maps those contrasting roles onto its systems metaphor without claiming an engineered purpose.

4. The Imaginary 7-Stage Cosmic Deployment Breakdown

The following breakdown uses the pipeline metaphor to organise the major scientific stages that made life possible. *It is imaginary 7-Stage pipeline but in Cosmic Architecture, there could be unimaginable stages of deployments.* Its purpose is explanatory: to show sequence, dependency and accumulation across deep time without implying that astronomy has discovered literal cosmic software jobs.

Stage 1: Core Infrastructure Provisioning (0–380,000 Years After the Big Bang)

- **Repository Source:** core-laws/spacetime-manifold
- **Semantic Version Tag:** v1.0.0-INITIAL-RELEASE
- **Astrophysical Event:** The Big Bang initializes spacetime. Rapid cosmic inflation expands the spatial grid, subatomic particles form simple hydrogen and helium, and light is released as the Cosmic Microwave Background (CMB).
- **Pipeline Interpretation:** The metaphor treats the emergence of spacetime, constants and quantum fields as the first enabling conditions of the observable universe. Scientifically, these are

described by cosmology and physics; the pipeline language is used only to organise the sequence.

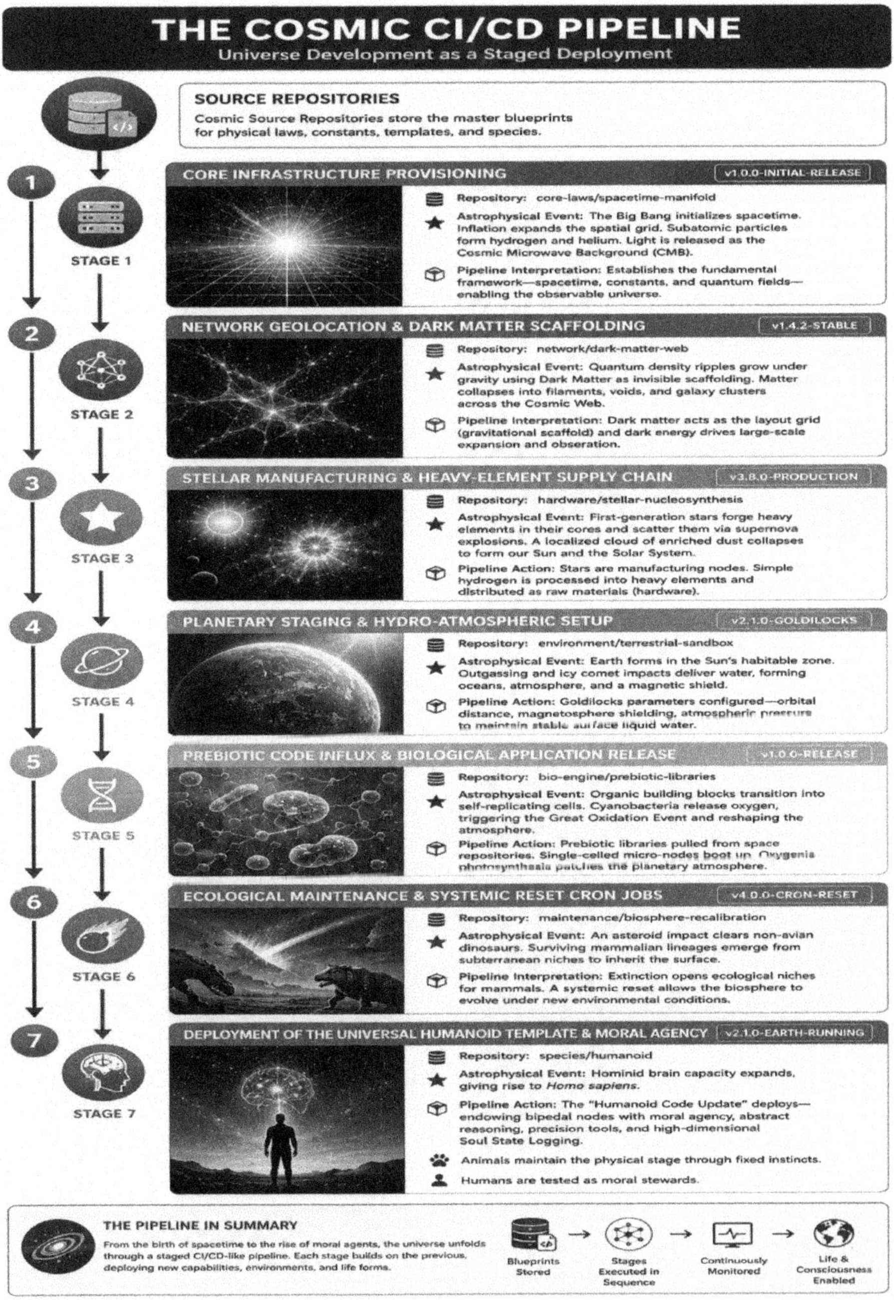
THE COSMIC CI/CD PIPELINE
Universe Development as a Staged Deployment
SOURCE REPOSITORIES
Cosmic Source Repositories store the master blueprints for physical laws, constants, templates, and species.
1
STAGE 1
CORE INFRASTRUCTURE PROVISIONING
v1.0.0-INITIAL-RELEASE
Repository: core-laws/spacetime-manifold
Astrophysical Event: The Big Bang initializes spacetime. Inflation expands the spatial grid. Subatomic particles form hydrogen and helium. Light is released as the Cosmic Microwave Background (CMB).
Pipeline Interpretation: Establishes the fundamental framework—spacetime, constants, and quantum fields—enabling the observable universe.
2
STAGE 2
NETWORK GEOLOCATION & DARK MATTER SCAFFOLDING
v1.4.2-STABLE
Repository: network/dark-matter-web
Astrophysical Event: Quantum density ripples grow under gravity using Dark Matter as invisible scaffolding. Matter collapses into filaments, voids, and galaxy clusters across the Cosmic Web.
Pipeline Interpretation: Dark matter acts as the layout grid (gravitational scaffold) and dark energy drives large-scale expansion and observation.
3
STAGE 3
STELLAR MANUFACTURING & HEAVY-ELEMENT SUPPLY CHAIN
v3.8.0-PRODUCTION
Repository: hardware/stellar-nucleosynthesis
Astrophysical Event: First-generation stars forge heavy elements in their cores and scatter them via supernova explosions. A localized cloud of enriched dust collapses to form our Sun and the Solar System.
Pipeline Action: Stars are manufacturing nodes. Simple hydrogen is processed into heavy elements and distributed as raw materials (hardware).
4
STAGE 4
PLANETARY STAGING & HYDRO-ATMOSPHERIC SETUP
v2.1.0-GOLDILOCKS
Repository: environment/terrestrial-sandbox
Astrophysical Event: Earth forms in the Sun's habitable zone. Outgassing and icy comet impacts deliver water, forming oceans, atmosphere, and a magnetic shield.
Pipeline Action: Goldilocks parameters configured—orbital distance, magnetosphere shielding, atmospheric pressure to maintain stable surface liquid water.
5
STAGE 5
PREBIOTIC CODE INFLUX & BIOLOGICAL APPLICATION RELEASE
v1.0.0-RELEASE
Repository: bio-engine/prebiotic-libraries
Astrophysical Event: Organic building blocks transition into self-replicating cells. Cyanobacteria release oxygen, triggering the Great Oxidation Event and reshaping the atmosphere.
Pipeline Action: Prebiotic libraries pulled from space repositories. Single-celled micro-nodes boot up. Oxygenic photosynthesis patches the planetary atmosphere.
6
STAGE 6
ECOLOGICAL MAINTENANCE & SYSTEMIC RESET CRON JOBS
v4.0.0-CRON-RESET
Repository: maintenance/biosphere-recalibration
Astrophysical Event: An asteroid impact clears non-avian dinosaurs. Surviving mammalian lineages emerge from subterranean niches to inherit the surface.
Pipeline Interpretation: Extinction opens ecological niches for mammals. A systemic reset allows the biosphere to evolve under new environmental conditions.
7
STAGE 7
DEPLOYMENT OF THE UNIVERSAL HUMANOID TEMPLATE & MORAL AGENCY
v2.1.0-EARTH-RUNNING
Repository: species/humanoid
Astrophysical Event: Hominid brain capacity expands, giving rise to Homo sapiens.
Pipeline Action: The "Humanoid Code Update" deploys—endowing bipedal nodes with moral agency, abstract reasoning, precision tools, and high-dimensional Soul State Logging.
Animals maintain the physical stage through fixed instincts.
Humans are tested as moral stewards.
THE PIPELINE IN SUMMARY
From the birth of spacetime to the rise of moral agents, the universe unfolds through a staged CI/CD-like pipeline. Each stage builds on the previous, deploying new capabilities, environments, and life forms.
Blueprints Stored
Stages Executed in Sequence
Continuously Monitored
Life & Consciousness Enabled

Stage 2: Network Geolocation & Dark Matter Scaffolding(~100 million–9.2 billion years after Big Bang)

- **Repository Source:** network/dark-matter-web
- **Semantic Version Tag:** v1.4.2-STABLE
- **Astrophysical Event:** Quantum density ripples grow under gravity using Dark Matter as invisible scaffolding. Matter collapses into long filaments, voids, and galaxy clusters across the Cosmic Web.
- **Pipeline Interpretation:** The metaphor describes dark matter as a layout-like gravitational scaffold and dark energy as a large-scale expansion factor. These terms organise the chapter's analogy; they do not identify dark matter or dark energy as engineered software components.

Stage 3: Stellar Manufacturing & Heavy-Element Supply Chain (~9.2 billion years after Big Bang / 4.6 billion years ago)

- **Repository Source:** hardware/stellar nucleosynthesis
- **Semantic Version Tag:** v3.8.0-PRODUCTION
- **Astrophysical Event:** First-generation stars forge heavy elements inside their cores and scatter them across space via supernova explosions. A localized cloud of enriched dust collapses to form our Sun, Earth, and Solar System.
- **Pipeline Action:** Stars operate as automated manufacturing nodes. Simple hydrogen is processed into carbon, oxygen, silicon, and iron. Supernovae act as distribution triggers, supplying raw hardware ingredients to stage rocky worlds.

Stage 4: Planetary Staging & Hydro-Atmospheric Setup (~9.8 billion years after Big Bang / ~4.0 billion years ago)

- **Repository Source:** environment/terrestrial-sandbox
- **Semantic Version Tag:** v2.1.0-GOLDILOCKS
- **Astrophysical Event:** Earth forms in the Sun's habitable zone. Mantle outgassing and icy comet impacts deliver water, forming oceans, a protective atmosphere, and a core-driven magnetic shield.

- **Pipeline Action:** The pipeline configures Goldilocks parameters. It balances orbital distance, magnetosphere deflector shielding, and atmospheric pressure to maintain stable surface liquid water.

Stage 5: Prebiotic Code Influx & Biological Application Release (~13.2 billion years after Big Bang / ~600 million years ago)

- **Repository Source:** bio-engine/prebiotic-libraries
- **Semantic Version Tag:** v1.0.0-RELEASE
- **Astrophysical Event:** Organic building blocks transition into self-replicating cells. Cyanobacteria release oxygen, triggering the Great Oxidation Event and reshaping Earth's atmosphere.
- **Pipeline Action:** The pipeline pulls prebiotic libraries from space repositories. Single-celled micro-nodes boot up. Automated patches (oxygenic photosynthesis) update the planetary atmosphere to support complex life.

Stage 6: Ecological Maintenance & Systemic Reset Cron Jobs (~13.735–13.8 billion years after Big Bang / ~65 million years ago to present)

- **Repository Source:** maintenance/biosphere-recalibration
- **Semantic Version Tag:** v4.0.0-CRON-RESET
- **Astrophysical Event:** An asteroid impact clears non-avian dinosaurs. Surviving mammalian lineages emerge from subterranean niches to inherit the surface.
- **Pipeline Interpretation:** The extinction event can be understood scientifically as a catastrophic environmental disruption that opened ecological niches for mammals. The systems metaphor calls this a reset, but it should not imply that mass extinction was deliberately scheduled or morally justified.

Stage 7: Deployment of the Recurring Humanoid Template & Moral Agency (Present Epoch)

- **Repository Source:** species/humanoid
- **Semantic Version Tag:** v2.1.0-EARTH-RUNNING

- **Astrophysical Event:** Hominid brain capacity expands, giving rise to *Homo sapiens.*
- **Pipeline Action:** The pipeline deploys the "Humanoid Code Update"—endowing bipedal nodes with unbound moral agency, abstract reasoning, precision tools, and high-dimensional Soul State Logging. Animals maintain the physical stage through fixed instincts, while humans are tested as moral stewards.

5. Cosmic Pipeline Summary Table

Pipeline Stage	**Astrophysical Event & Timeline**	**Systems Engineering & Repository Equivalent**
Stage 1: Core Provisioning	Big Bang (t = 0) to CMB (yrs)	**Infrastructure as Code (core-laws: v1.0.0)**: Spacetime grid, light speed, and quantum rules provisioned.
Stage 2: Network Geolocation	Cosmic web forms	**Network Routing (dark-matter: v1.4.2)**: Dark Matter grid deployed; Dark Energy configured for zone isolation.
Stage 3: Hardware Staging	Solar system forms	**Hardware Enrichment (stellar-forge: v3.8.0)**: Stars process hydrogen into carbon, oxygen, and iron via supernovae.
Stage 4: Planetary Staging	Earth oceans stabilize	**Environment Staging (terrestrial: v2.1.0)**: Magnetosphere and Goldilocks liquid water parameters initialized.

Pipeline Stage	**Astrophysical Event & Timeline**	**Systems Engineering & Repository Equivalent**
Stage 5: App Deployment	First cells & cyanobacteria	**Initial Software Release (bio-engine: v1.0.0)**: Prebiotic cometary delivery & atmospheric OS patch.
Stage 6: System Maintenance	End-Cretaceous impact	**Scheduled Cron Job (maintenance: v4.0.0)**: Dominant macro-nodes pruned; mammalian sandbox unlocked.
Stage 7: Moral Sandbox	Present Epoch: Humans on Earth	**Moral Agency Release (species/humanoid: v2.1.0)**: Hominid code update deployed; moral choice and soul logging active.

6. The Micro-Pipeline: How One DNA Blueprint Produces Different Cells

The cosmic pipeline analogy can also be applied at a much smaller scale: **inside the human body**.

A human body contains trillions of cells. A brain cell looks and behaves very differently from a heart cell, a skin cell, or a liver cell. Yet most of these cells contain essentially the **same DNA**.

So how can the same genetic blueprint produce completely different types of cells?

The answer is that **not every gene is active in every cell**.

Think of DNA as an enormous instruction library containing thousands of biological instructions. A heart cell does not need to use every instruction required by a brain cell, and a brain cell does not need to use every instruction required by a liver cell. Instead, cells use regulatory mechanisms to control which genes are switched **on**, which are switched **off**, and how strongly particular genes are used.

This is called **gene regulation**.

In systems-engineering language, the body therefore does not need a separate DNA installation for every type of cell. It can reuse the same underlying genome while activating different parts of it in different biological environments.

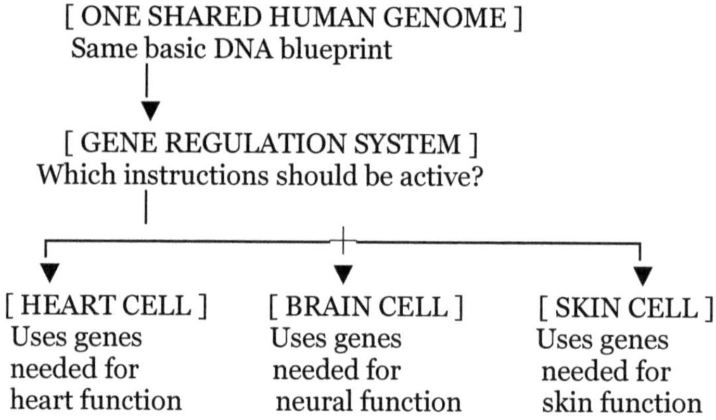

A simple analogy is a **large software platform containing many available modules**. Every installation may begin with the same underlying codebase, but its configuration determines which modules actually run.

The biological reality is much more complicated than software, but the analogy helps explain an important principle:

Same DNA does not mean every cell executes the same biological instructions.

Evolutionary Adaptation and Systemic Reuse

The same general idea of **reusing an existing biological system rather than rebuilding everything from scratch** appears throughout evolution.

Evolution usually works by modifying structures and regulatory systems inherited from earlier generations. Changes in when, where, and how strongly genes are expressed can alter biological characteristics without requiring an entirely new genome.

Hormones provide an easy example. Hormones are **chemical signals that travel through the body**. Cells contain receptors that can respond to particular hormonal signals. During development, differences in hormone levels and tissue responses can influence multiple characteristics at the same time, including muscle development, body-hair patterns, fat distribution, and aspects of reproductive development.

Human populations also show adaptations associated with different environments. **Skin pigmentation**, for example, is influenced by genes affecting melanin production, while some populations living at high altitude have inherited biological adaptations that help their body's function where oxygen availability is lower.

These examples are not produced by one universal biological "configuration switch," and the mechanisms differ considerably. The broader principle is simply that **evolution often modifies and regulates existing biological machinery rather than constructing an entirely new organism from nothing.**

In the language of the *Cosmic Architecture*, we can therefore imagine:

Shared biological platform → inherited genetic variation → regulatory differences → different biological outcomes

This is why the software analogy is useful. A single underlying biological architecture can support enormous variation.

Within the speculative language of this book, humanity can therefore be pictured as a highly flexible biological platform—humanoid: v2.1.0—capable of producing many physical variations while retaining the same fundamental human architecture.

Chapter 10

Fine-Tuning and System Configuration – The Cosmic Parameters

When we look deeply into the laws of physics, we encounter one of the most profound mysteries in science: the question of fine-tuning.

To put it simply, our universe appears to run on a highly specific set of rules and numerical values. If the fundamental features of our universe—such as the strength of gravity or the mass of an electron—were altered even slightly, the entire system would fail. Long-lived stars could not form, complex chemistry would be impossible, and life as we know it could never exist.

If we view the cosmos as a highly advanced architectural framework, these physical constants are the system configuration parameters that allow the universe to successfully boot up and host conscious life.

However, recognizing this incredible precision does not automatically mean science has mathematically "proved" a designer in a laboratory. Instead, scientists and philosophers debate several possible explanations for why our universe is so perfectly balanced. To truly understand the architecture of reality, we must examine these possibilities.

1. The Anthropic Principle: The "Selection Effect"

One of the most common scientific explanations for fine-tuning is the Anthropic Principle. This principle states a very simple, logical fact: observers (human beings) can only find themselves existing in a universe that has the right conditions to support observers.
In other words, we shouldn't be surprised that the universe looks perfectly tuned for life, because if it weren't perfectly tuned, we wouldn't be here to notice it!

While this statement is logically accurate, it is incredibly modest. It describes a selection effect, but it does not actually explain why such perfect conditions exist in the first place.

Imagine surviving a firing squad of 100 trained marksmen who all somehow miss you. Saying, "Well, I shouldn't be surprised they missed, because if they hit me, I wouldn't be alive to think about it," is logically true. But it completely ignores the deeper question:

Why did they all miss? Was it rigged?

The Anthropic Principle tells us that we are in a working system. It does not, by itself, settle why such conditions are possible, whether they arise from necessity, chance, a multiverse, or design. That unresolved space is where philosophical interpretation begins.

2. The Multiverse vs. Iterative Deployments

To explain away the miraculous nature of our fine-tuned universe without invoking a Creator, some physicists point to the Multiverse. They suggest that perhaps there are infinite universes, all bubbling into existence with different, random conditions and physical laws. In an infinite lottery, eventually, one universe will hit the exact winning combination for life—and that just happens to be the one we live in.

However, within *Cosmic Architecture framework*, we can look at multiple universes differently. Rather than a chaotic, random foam of infinite mistakes, a multiverse can be understood as sequential, iterative deployments:

- **Cross-Universal Bootstrapping:** Instead of random chance, universes may be linked. A previous, older universe (Universe A) may have run its course, generating data, structural lessons, and basic evolutionary blueprints.
- **Upgraded Configurations:** Those lessons are then preserved and used to establish the "initial conditions" and fine-tuned parameters for the next deployment (Universe B).

This is one possible way to imagine a multiverse as iterative rather than purely random. The book offers it as a structured speculative model, not as a replacement for scientific multiverse proposals or as a proven description of reality.

3. Deeper Laws and the Question of Design

Some scientists hope that one day we will discover a "deeper law" of physics—a single, unifying mathematical equation that proves the universe had to be this exact way and couldn't possibly exist in any other configuration. But even if we find that master equation, the fundamental architectural question remains: *Why does reality contain a master equation capable of producing such perfectly ordered, life-sustaining structures at all?*

The Cosmic Architecture framework keeps these possibilities in perspective:

- **Physics maps the parameters:** It measures the exact numbers that make life possible.
- **Philosophy interprets the meaning:** It suggests that a universe governed by hyper-precise, life-enabling mathematics points strongly to an underlying intelligence.

The fine-tuning of the universe reveals a breathtaking precision that invites interpretation. Some readers may see undiscovered physical necessity; others may see multiverse selection; still others may see deliberate configuration by a Supreme Cosmic Mind. The *Cosmic Architecture* does not force a single conclusion here. It asks the reader to consider why reality is intelligible, ordered and life-permitting at all.

Chapter 11

High-Dimensional Oversight, Human Free Will, and the Stress Tests of Time

1. The High-Resolution Cosmic Ledger (The Logging Subsystem)

- **Religious Parallel:** Jainism, Hinduism, and other traditions speak of divine surveillance ("God is always watching") and the meticulous record of Karma.
- **System Architecture Equivalent:** The model imagines intentional actions, choices and state changes as if they were recorded in an unalterable cosmic log. This is a metaphysical analogy for moral consequence, not a claim that a measurable database has been scientifically detected.
- **Reformation & Cloning:** When a node (soul/body) closes, its karmic state vector is committed to the database. When the node is re-spawned (reformed), those state vectors are cloned into the new instance, carrying over past parameters.

2. Circuit Breakers: Corrective Action Before Failure

- **Religious Parallel:** Interventions where a person is mysteriously protected from imminent harm or guided away from a fatal path.
- **System Architecture Equivalent:** Site Reliability Engineering (SRE) uses **automated circuit breakers** and pre-emptive triggers.
- **The Mechanism:** When a localized node is about to suffer catastrophic, non-scheduled failure due to an external anomaly, the cosmic management system triggers a rapid intervention (a threshold safeguard) before the system breaks.

3. The "Hitler Dilemma": Why Allow Malicious Actors?

- **The Classic Question:** *"If the cosmos is logged and guarded, why permit dictators, warlords, and atrocious crimes to harm millions?"*
- **The Systemic Answer (Penetration Testing):**
    - Dark actors and severe societal failures act as **Cosmic Penetration Tests (Red Teaming)** for human civilization.
    - They expose critical vulnerabilities in human morality, governance, and law.
    - If a system never experiences stress tests, it cannot strengthen its security posture or evolve. Bad actors show humanity its worst vulnerabilities, forcing civilization to write better laws, develop higher moral standards, and build preventive safeguards.

4. The Cosmic Non-Interference Policy & Project Discard Threshold

- **The Boundary Rule:** The Cosmos enforces a strict **Non-Interference Policy**. It injects guidance from time to time (Avatars, enlightened teachers, moral patches) to illuminate the blueprint, but it **never forces** humanity to fix its code. Humanity must voluntarily implement the patches.
- **The Danger of Ignored Recommendations:** Just like an IT security team that repeatedly ignores security audit reports until a fatal ransomware attack wipes out the enterprise, a civilization that continually ignores cosmic lessons (choosing war, mass destruction, and moral decay) risks total system failure.
- **The Project Discard Protocol:** Humanity is an active development project. If the project repeatedly fails its evaluation gates despite centuries of lessons, the Cosmos does not manually destroy it out of anger—the system simply lets the project terminate itself under its own broken logic (a discarded project).

5. 3,000 Years of System Evolution: Toward an Advanced Civilization

- **Historical Evidence of Moral Learning:** Looking back across thousands of years—from tribal conquest, persecution and mass violence to modern human rights language, international law and public moral accountability—suggests that humanity can learn from severe failure. This does not prove that suffering is designed as a test; it shows that societies can convert catastrophe into reform when they choose to learn.
- **The Current Milestone:** Ongoing wars, nuclear threats and ecological risks may be understood as severe civilisational stress points. Whether humanity responds with restraint, cooperation and moral maturity remains an open question. The model frames this as a test of collective learning, not as a prediction that history must end in either salvation or collapse.

PART II — SYSTEM ARCHITECTURE TAKEAWAYS

The Physical Sandbox & Hardware

1. Spacetime as a Metric and Causal Light Cones

Physical space is not an empty void but a dynamic four-dimensional continuous manifold. The invariant speed of light (**c**) establishes the fundamental boundary of causality, protecting the timeline from spacelike paradoxes and preventing unauthorized state rollbacks within the physical sandbox.

2. Automated Cosmic Deployment Pipeline (CI/CD)

The 13.8-billion-year physical history of the universe can be understood as an automated, staged deployment pipeline. The process proceeds from core spacetime provisioning to dark matter scaffolding, stellar element forging, planetary hydro-atmospheric configuration, and ultimately the release of biological life as a higher-level application layer.

3. Balanced Infrastructure: Dark Matter and Dark Energy

Dark matter contributes gravitationally to the formation of the Cosmic Web and the growth of galaxies and clusters. Dark energy is associated with the accelerated expansion of the universe. The Cosmic Architecture interprets these established roles metaphorically as structural scaffolding and large-scale separation, while treating any governance or tenant-domain purpose as speculation.

4. Fine-Tuning and Stress-Test Governance

Physical constants represent extraordinarily precise configuration values that allow complex chemistry, stable stars, planetary habitability, and conscious life to emerge. Within this framework, free agency is permitted under a Non-Interference Policy, while historical catastrophes, conflicts, and civilizational failures function as large-scale stress tests that compel societies to develop stronger institutions, improved laws, and greater moral maturity.

Part II Core Principle

Within the systems analogy, the physical universe can be pictured as an execution environment: spacetime supplies the geometric framework, dark matter contributes to large-scale structure, dark energy is associated with accelerated expansion, and physical parameters help determine which kinds of complexity can arise. The engineering language is interpretive scaffolding rather than evidence that these components were deliberately configured.

PART III - LIFE BEFORE AND AFTER LIFE

Looking Back:

In **Part II**, we mapped the physical stage of reality. We examined how spacetime metrics, quantum fields, dark matter scaffolding, and stellar nucleosynthesis operate as an automated deployment pipeline to forge rocky, water-bearing worlds like Earth. We established the hard causal boundaries of physics that any serious philosophical model must respect.

Moving Forward:

Having built the physical stage, **Part III** explores the transition from inert physical hardware to dynamic, conscious software. We step across the threshold of abiogenesis—examining chemistry before biology, the delivery of organic libraries via space vectors, and the emergence of cumulative cultural memory. We then tackle the deepest questions of existence: moral agency, the generation of the lifetime state log, the mechanics of post-shutdown state cloning (re-embodiment), and how conscious entities graduate into multi-tiered governance realms. Part III bridges planetary science with the enduring journey of the soul record.

Chapter 12

Chemistry Before Biology – Bootstrapping the Cosmic Engine

Evolution provides an extraordinarily powerful account of how living populations adapt and diversify once reproduction, inheritance, and variation are already active. It explains how a finch's beak changes over generations or how mammals adapted to new landscapes. However, evolutionary theory naturally begins *after* life is already up and running. It does not by itself answer the deeper, preceding question: How did the very first evolvable system spring from non-living matter?

This threshold between inert geology and self-sustaining biology is known as **abiogenesis**. Where standard biology sees an unsolved chemical puzzle, the framework of *Cosmic Architecture* recognizes a fundamental system initialization—the delicate phase were raw chemical hardware transitions into dynamic, self-executing software.

The Pre-Biological Sandbox

To understand how life began, science explores how non-living chemistry could naturally assemble systems capable of storing information, harvesting energy, and replicating. Researchers investigate early planetary environments, self-organizing fatty molecules, mineral surfaces that act as templates, and networks of chemical reactions driven by thermal energy gradients. Yet, despite decades of research, no single, complete pathway from simple geochemistry to the first functioning cell has been established.

A living cell is not merely a random mixture of chemicals. It is a highly integrated architectural package requiring four distinct system components:

1. **Container Boundaries:** Protective membranes that separate the internal system from the external environment.
2. **Energy Processing:** Active metabolic drivers that power continuous internal processes.

3. **Information Storage:** Code-like molecules (such as RNA or DNA) that hold operational instructions.
4. **Replication Protocols:** Reliable mechanisms to copy and pass those instructions forward.

Scientists have discovered that many of the building blocks of life can form naturally.

Fatty molecules can join to create tiny bubble-like structures like cell walls.

Small chemical molecules can link together to form longer chains, and natural energy sources, such as hot underwater vents, can power chemical reactions.

The biggest unanswered question is not whether these individual parts can form, but how they all came together at the same time to create the first living cell—a system capable of protecting itself, using energy, storing information, and making copies of itself.

The Deep-Space Transport Question

This chemical foundation directly impacts how we evaluate theories like panspermia—the idea that life arrived on Earth from comets or interstellar debris. Asking whether a fully formed, modern bacterium could have emerged out of the deep freeze of space begins far too late in the narrative. Deep space lacks the stable atmospheric pressure, liquid water, and thermal energy loops required to support complex, active biology.

However, rich prebiotic chemistry *does* occur in space. Comets formed in the cold outer reaches of our young solar system contain surprisingly complex organic molecules, water ice, and mineral dust. The true architectural question is not whether living organisms were floating freely in the void waiting to collide with Earth, but whether the fundamental building blocks—the raw molecular libraries—were forged across cosmic distances and delivered to planetary surfaces.

A comet carrying amino acids or complex carbon molecules is not carrying life itself; it is delivering essential raw materials. If living micro-organisms were ever discovered inside ancient comet ice, we would still face a critical system audit: *Did that life boot up inside the frozen comet, was it picked up from an earlier planetary environment, or was the sample contaminated upon collection?* Each scenario carries a vastly different burden of proof.

Natural Law as the System Runtime

In scientific discussions, the word *"Nature"* is simply shorthand for physical and biological processes—environmental pressures, chemical laws, energy flows, and heredity. Natural selection itself is not a conscious entity sitting on a throne making deliberate choices. It is an unfeeling, automated filter: the varying rates of survival and reproduction acting upon small differences in living organisms.

Cosmic Architecture does not reject these natural processes. Instead, it views them as the **runtime environment** established by the cosmos.

When a computer executes an algorithm, the algorithm runs within an environment configured with rules, limits and capacities. In a similar way, chemistry becomes biologically interesting only because the universe permits stable atoms, bonds, energy gradients and self-organising molecular structures. The Cosmic Architecture interprets those enabling conditions as suggestive of deeper order, while recognising that abiogenesis remains an active scientific research problem rather than a closed proof of design.

Chapter 13

Panspermia as Transport, Not Origin – The Migration Fallacy

When faced with the sheer chemical improbability of life spontaneously booting up on early Earth, some scientific theories look upward to the stars. They propose **Panspermia**—the idea that life did not originate on Earth at all, but was carried here from deep space aboard asteroids, meteorites, or icy comets.

While this hypothesis offers a fascinating look at how physical matter moves across the cosmos, it frequently creates a massive, logical illusion. It tempts us to mistake **transportation** for **creation**.

To understand the true architecture of the cosmos, we must clearly separate the mechanism of moving life across space from the fundamental question of how life came to exist in the first place.

The Five Steps of the Transport Pipeline

Panspermia is essentially an interstellar supply chain. To work, it relies on a five-stage sequence:

1. **Origin:** Life must first come into being on a primary planet or celestial body.
2. **Enclosure:** Living organisms must be sheltered inside protective rock, ice, or deep porous minerals.
3. **Ejection & Transport:** A violent asteroid impact blasts those rocks into space, launching them on long transit trajectories through the vacuum.
4. **Shielded Survival:** The microscopic passengers must survive centuries of freezing absolute zero, harsh cosmic radiation, and deep-space vacuum.
5. **Arrival & Deployment:** The rock enters a new planet's atmosphere, survives the fiery heat of descent, and impacts a world capable of hosting life.

When mapped out, the logical flaw in relying on Panspermia as an ultimate answer becomes obvious. Steps two through five detail *logistics*—how a package is sealed, shipped, insulated, and delivered. But none of those steps address Step One.

Panspermia explains migration, not origin. Moving a living cell across space explains how a passenger reached a new destination, but it tells us nothing about how the passenger was born.

Moving Data vs. Writing the Code

In system engineering, moving a data backup from a local server to a remote cloud repository shifts where the information lives, but it does not explain who authored the original application logic.

Similarly, if life were discovered inside an ancient meteorite, it would be an extraordinary astronomical finding, but it would not solve the mystery of life's origin. It would simply push the origin question back in time and transfer it to a different location in the universe.

We would still have to ask: *Where did the original planet get its life? How did those first self-replicating strands assemble themselves out of dead chemistry?*

Transport answers *"How did it get here?"*

Origin answers *"How did it first exist?"*

The Core Architectural Truths

To keep our cosmic framework grounded, we must hold onto four distinct boundaries:

- **Transport is Not Origin:** The vessel comes *after* the cell. A comet or meteorite can only carry life if life already exists to be loaded onto it.

- **Evolution is Not a Ladder:** Evolution splits populations into diverging branches over time. A chimpanzee and a human share a distant common ancestor, just as modern bacteria and complex organisms branch from early nodes. Evolution modifies existing systems; it does not invent the initial system logic out of nothing.
- **Chemistry is Not Biology:** Space is filled with organic, carbon-based molecules, but a list of chemical ingredients is not a living organism. Life requires organized, self-executing system structures.
- **Movement Explains Location, Not Purpose:** Migrating life across planetary systems demonstrates the vast reach of cosmic hardware, but the ultimate question remains untouched: *The deeper question would still remain: how did the first life-producing informational organization arise anywhere in the cosmos?*

By understanding that panspermia concerns delivery rather than ultimate origin, we avoid mistaking movement for explanation. The cosmos may distribute organic molecules across vast distances, and perhaps even biological material under rare conditions, but the deeper question of how life first becomes self-sustaining remains open. In this book, that open question becomes a doorway into the larger architecture rather than a conclusion already proven.

Chapter 14

Evolution, Consciousness, and the Emergence of Intelligence

Standard biology teaches us a fundamental truth about nature: biological evolution does not keep a hidden master blueprint in a drawer, nor does it work toward human beings as a predetermined final product.

When an asteroid wiped out the non-avian dinosaurs 66 million years ago, nature did not reload an earlier backup file. Small nocturnal mammals survived simply because their small size, burrowing habits, and lower energy needs were advantageous during a global environmental catastrophe. Over tens of millions of years, those surviving lineages branched out in countless unpredictable directions. Humans represent one late branch on that tree, not a forced destination written at the beginning.

[Mass Extinction Event] —▶ Unplanned Environmental Shift
↓
[Surviving Lineages] —▶ Branching Variations (No Master Blueprint)
↓
[Human Emergence] —▶ Late Branch / High-Energy Cost Mutation

This raises a crucial question for our framework: *If local Earth evolution as per Cosmic Architecture model operates through natural selection without hand-crafting individual species, how does the humanoid form emerge across the multiverse?*

1. The Seed and the Staging Ground: Cosmic Delivery of the Bio-Code

To understand how the humanoid branch emerged on Earth, we must bridge evolutionary biology with cosmic delivery mechanisms.

In standard cosmology, early Earth was a hot, sterile planetary engine. Yet, within a few hundred million years of cooling, life booted up. *How did raw geochemistry assemble the complex information-bearing code required for cellular life?*

[COMETARY DELIVERY] ──▶ Prebiotic Organic Libraries (Amino Acids, Nucleobases)

↓

[EARTH ENVIRONMENT] ──▶ Optimal Water Sandbox & Energy Gradients

↓

[HUMANOID BRANCH] ──▶ Expression of Shared Multiverse Genetic Code

1. **Prebiotic Delivery via Comets:** Deep space is not an empty void; it is a chemical factory. Comets, icy meteorites, and carbonaceous asteroids traveling through interstellar space carry complex organic molecules—amino acids, nucleobases, and sugar precursors. Asteroids, carbonaceous meteorites, and perhaps comets delivered organic compounds to the early Earth. The scale and timing of a distinct Late Heavy Bombardment remain debated, and these bodies did not deliver known living organisms. Within the book's analogy, their organic chemistry can be pictured as a transported library of prebiotic ingredients rather than a finished biological program.
2. **Earth as the Perfect Staging Ecosystem:** Earth provided the ultimate liquid water sandbox, thermal heat loops, and magnetic protection. The delivered organic code combined with Earth's unique environment, initializing the first self-replicating cells.
3. **The Humanoid Possibility:** The book imagines that bipedal, tool-using, socially intelligent beings may represent one recurring pathway by which moral agency can emerge under suitable conditions. This is a speculative pattern, not a biological conclusion that humanoids must appear in every universe.

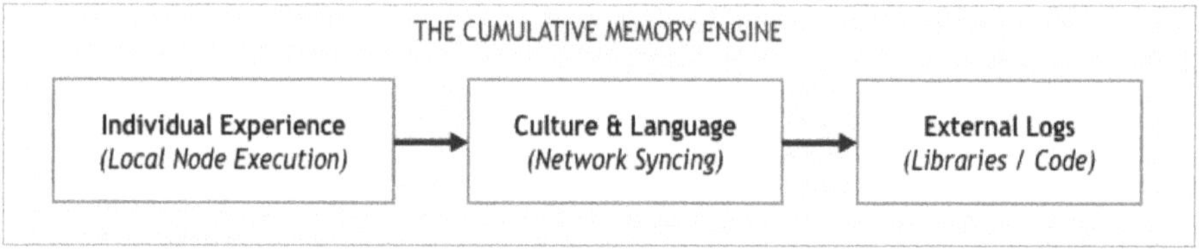

2. The Cost and Surprise of Consciousness

From a purely evolutionary standpoint, high intelligence is not automatically a superior survival strategy. A human brain consumes nearly 20% of the body's energy while making up only 2% of its mass. It takes decades to mature, creates significant birth risks, and requires enormous resources. Bacteria, ants, and sharks have thrived for hundreds of millions of years without ever writing books, building cities, or contemplating the cosmos.

Yet, human intelligence crossed a critical threshold through cumulative culture. Unlike animals whose learning remains largely limited to immediate survival instincts, humans developed language, storytelling, writing, libraries, and digital networks.

Civilization itself acts as an external memory architecture. It allows one generation to preserve its discoveries and pass them forward so the next generation does not have to reinvent the wheel.

The Synthetic Memory Accelerator Civilization originally expanded its wetware ceiling through external logs—first language, then writing, then global digital networks. Today, artificial intelligence represents the latest leap in this external memory architecture.

Rather than functioning merely as passive storage, machine intelligence acts as a **semantic compiler** over human civilization's accumulated telemetry. It compresses billions of human texts, proofs, and scientific insights into high-dimensional latent spaces, enabling rapid cross-domain synthesis that would take a single biological node lifetimes to calculate. It is the externalization of pattern recognition—allowing human conscious nodes to operate at higher conceptual abstraction.

3. The Brain as an Adaptive Algorithm

In simple terms, an algorithm is a procedure—a structured series of steps that transforms an input into an output. A cooking recipe is an algorithm; so is the process a child uses to read a word.

Calling the human brain an algorithmic system does not mean it functions like a rigid computer. It means the brain dynamically updates its internal neural patterns as new experience arrives, continually refining how it notices, remembers, and decides.

Most learning is traceable—we practice, fail, compare, and improve. Yet human history is filled with moments of sudden, unexplainable insight: intuitive leaps, sudden scientific breakthroughs, and deep artistic inspiration that seem to arrive fully formed after long periods of mental incubation.

Standard psychology explains this through unconscious background processing. But our cosmic model asks an additional, complementary question:

Could sudden intuitive breakthroughs occur when a local brain-node connects to an information network larger than its immediate 3D hardware?

This question is explored in detail in **Chapter 20, "Biological Wetware Ceilings, Gifted Minds, and Over-the-Air Knowledge Infusions,"** where the Cosmic Architecture considers whether highly prepared or unusually receptive minds could function as catalyst nodes for sudden insight. The broader limits and future possibilities of human cognition are revisited in **Chapter 42, "The Human Brain, Knowledge Boundaries, and Future Intelligence."**

4. Preservation over Destruction: The Architectural Principle

When we observe the destruction in physical nature—star explosions, mass extinctions, and forgotten empires—it is easy to assume that consciousness and learning simply disappear without a trace when a life ends.

However, in system architecture and software engineering, preservation is a fundamental design principle. Engineers do not simply wipe a complex, valuable database or a mature software system because it is old. They refactor it, archive it, migrate its state vectors, or place it into maintenance mode to preserve its accumulated intelligence.

The *Cosmic Architecture* does not require evolution to have a rigid, hand-placed blueprint for every biological body. Instead, it establishes an environment where pre-delivered genetic possibilities can evolve, accumulate experience, and build culture.

Chapter 15

The Mechanics of the Soul Record, Cloning, and Multi-Tier Governance

1. Moral Agency and the Generation of the Lifetime State Log

In the language of ancient spiritual traditions, this entity has been called a **soul** or **Jiva**. In the systems language of this book, it is modelled as a **Conscious Node** or **Persistent State Vector**: a way of imagining how identity, memory, character and moral trajectory might persist beyond a single physical lifetime. This is a metaphysical definition within the book's framework, not an experimentally established data structure.

Within the speculative Cosmic Architecture model, to understand how a soul persists across physical reality, we must first address why human life generates a lasting footprint while ordinary animal life operates within localized biological loops. In the animal kingdom, behaviour is largely governed by evolved biological drives, environmental triggers, and survival instincts. A lion hunting prey or a bird building a nest is executing an evolved biological routine optimized for immediate survival.
Human beings developed an unusually powerful combination of self-reflection, abstract reasoning, long-term planning, and intentional moral choice. Other animals display varying degrees of cognition, social learning, empathy, and decision-making, so the distinction is not that humans alone possess all such capacities. The Cosmic Architecture focuses instead on the exceptional scale at which humans can reflect on ethical rules, build institutions, preserve knowledge, and deliberately choose against immediate self-interest.

When a person exercises moral agency—choosing honesty over deceit, compassion over cruelty, or self-sacrifice over selfish greed—the choice is not merely a transient chemical flare inside the physical brain. In the *Cosmic Architecture*, every intentional choice acts as an execution input, continuously updating a high-dimensional, path-dependent Soul State Record. The soul is not an abstract, featureless cloud; it is a living, dynamic, high-dimensional state record that logs the accumulated history, moral trajectories, intentional alignments, and spiritual maturity of a conscious entity across its physical existence.

2. The Process of Record Finalization, Cloning, and Re-embodiment

When a physical body reaches its biological limit and dies, the local physical node terminates, but the accumulated Soul State Record is preserved. The transition from death to the next physical iteration operates through a precise three-stage lifecycle process:

Log Closure and Audit: At the moment of physical death, the local lifetime log is closed. The net moral and spiritual state generated during that lifetime—the entity's accumulated Karma—is calculated as a path-dependent state vector.
State Cloning: The finalized state record is cloned into a new initialization template for the next cycle. This cloning process ensures that no experience, moral breakthrough, or unresolved debt is lost. The cloned record carries forward the exact state coordinates established in the previous life.

Karma-Driven Re-embodiment
Heavy / Unresolved State (Negative Karma): If the closed record carries significant unresolved corruption, harm, or moral degradation, the cloned record initializes under constrained, challenging physical conditions. These difficult conditions act as a natural recalibration environment, forcing the entity to learn discipline, empathy, and integrity through physical struggle.
Refined / Harmonious State (Positive Karma): If the closed record reflects constructive growth, selflessness, and moral clarity, the cloned record initializes under favourable, rewarding conditions, providing greater scope, tools, and opportunities for further spiritual development.
This process of closing, cloning and re-embodiment is presented as the book's metaphysical interpretation of reincarnation and moral continuity. It is not a demonstrated mechanism in contemporary science. Its purpose is to give traditional ideas of karma and rebirth a structured systems-language model that readers can test philosophically.

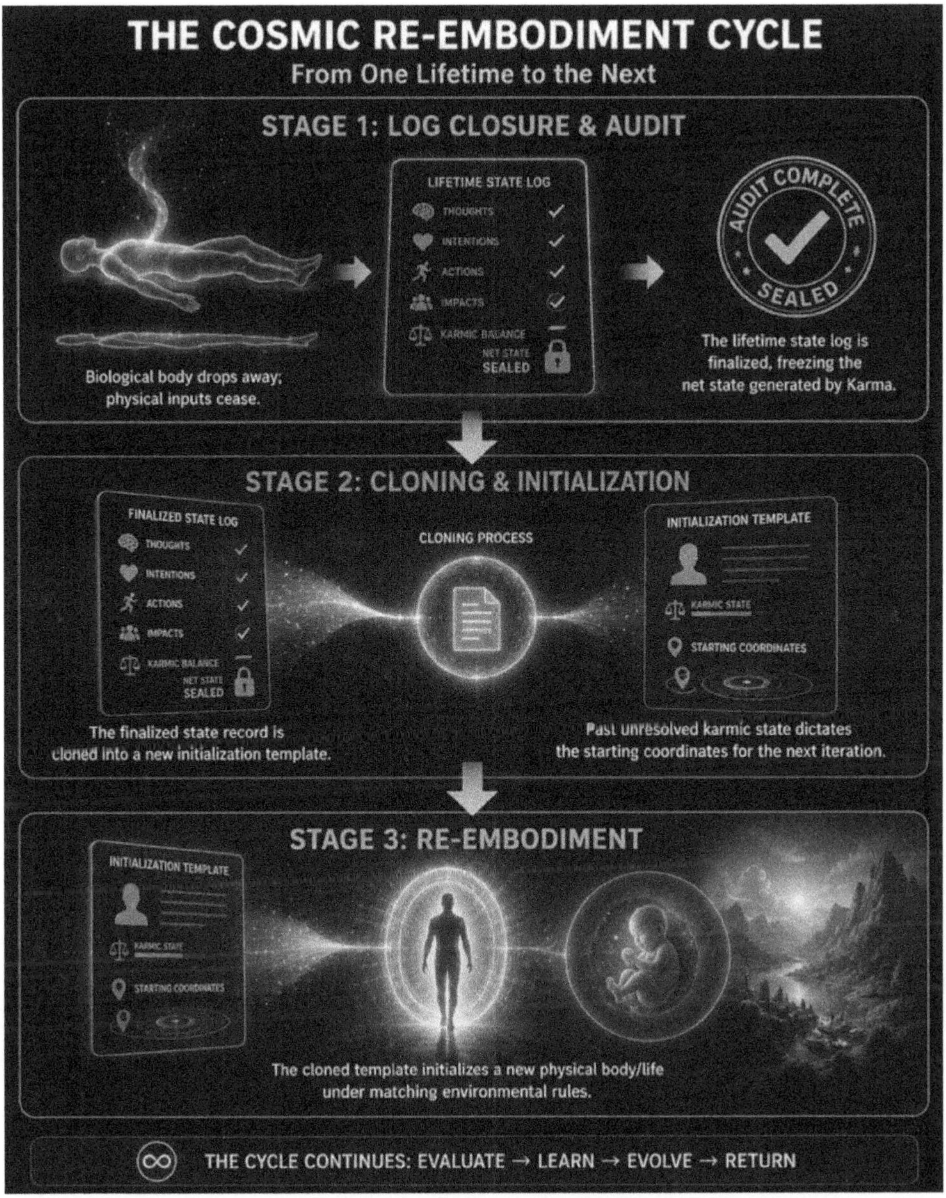
THE COSMIC RE-EMBODIMENT CYCLE
From One Lifetime to the Next
STAGE 1: LOG CLOSURE & AUDIT
LIFETIME STATE LOG
THOUGHTS
INTENTIONS
ACTIONS
IMPACTS
KARMIC BALANCE
NET STATE SEALED
AUDIT COMPLETE
SEALED
Biological body drops away; physical inputs cease.
The lifetime state log is finalized, freezing the net state generated by Karma.
STAGE 2: CLONING & INITIALIZATION
FINALIZED STATE LOG
THOUGHTS
INTENTIONS
ACTIONS
IMPACTS
KARMIC BALANCE
NET STATE SEALED
CLONING PROCESS
INITIALIZATION TEMPLATE
KARMIC STATE
STARTING COORDINATES
The finalized state record is cloned into a new initialization template.
Past unresolved karmic state dictates the starting coordinates for the next iteration.
STAGE 3: RE-EMBODIMENT
INITIALIZATION TEMPLATE
KARMIC STATE
STARTING COORDINATES
The cloned template initializes a new physical body/life under matching environmental rules.
THE CYCLE CONTINUES: EVALUATE → LEARN → EVOLVE → RETURN

3. The Non-Physical Governance Tier and Extended State Evaluation

Physical embodiment is not an endless, inescapable hamster wheel. It serves as a foundational training environment—a cosmic sandbox bounded by physical constraints, mortality, and material consequences.
When a soul record consistently refines its state across multiple physical lifetimes, mastering moral agency and selflessness, it reaches a critical threshold where physical embodiment is no longer required. The entity moves up into the Non-Physical Governance Tier.

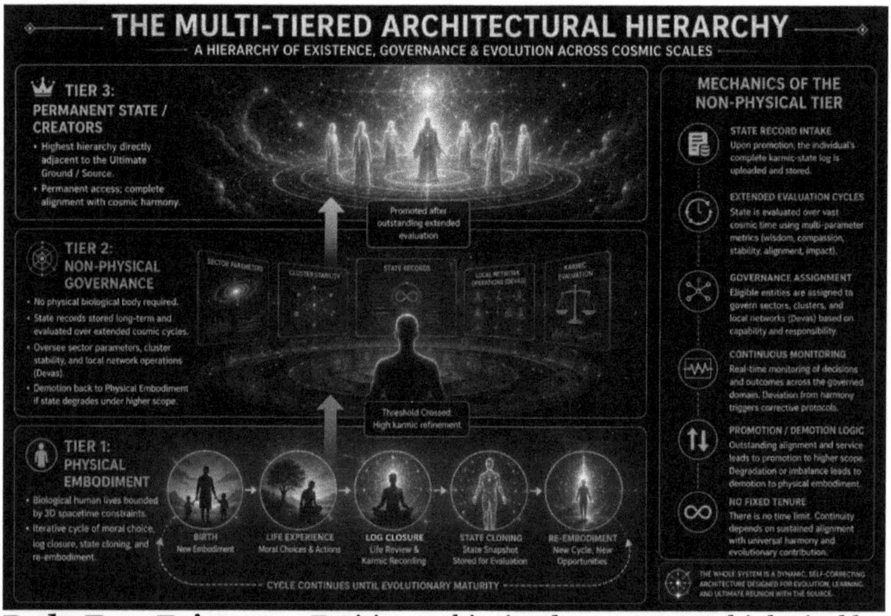

Body-Free Existence: Entities at this tier do not possess biological bodies made of atoms and flesh. They exist as high-dimensional, conscious state structures operating in parallel non-physical realms (what ancient traditions refer to as Devas or higher administrative beings).
Extended State Evaluation: In this tier, state records are not closed and cloned every few decades like human physical lives. Instead, records are maintained far longer and evaluated over extended cosmic cycles based

on how the entity exercises its broader awareness, administrative responsibility, and stewardship over lower physical clusters.

Promotion to the Permanent Tier: If an entity in the Non-Physical Governance Tier maintains exceptional integrity, purity, and alignment over these extended evaluation cycles, it reaches the ultimate threshold. Its record becomes permanent, and it is promoted to the highest hierarchy directly adjacent to the Ultimate Source / Creators. In this permanent state, the entity is completely free from death, decay, or forced recycling.

Demotion and Recalibration: Conversely, if an entity at the non-physical tier misuses its higher scope, accumulates spiritual pride, or allows its state record to degrade, the system triggers a recalibration. The entity is demoted back into physical embodiment—cloned back into a biological physical body on a world like Earth to re-learn fundamental lessons through material constraint and vulnerability.

4. Why This Complete System Matters

This state-cloning and multi-tier evaluation framework bridges ancient spiritual metaphysics (Karma, Reincarnation, Devas, Lokas, and Liberation) with modern systemic logic:

It Eliminates Cosmic Waste: Nothing you learn, endure, or achieve in a human lifetime is thrown away. Every choice contributes to the permanent evolutionary profile of your soul record.

Within the model, it offers an interpretation of human disparity: differences in talent, innate wisdom, moral maturity, or physical circumstances at birth can be understood as starting coordinates associated with a record carrying forward its previous state.

It Defines Purpose: Life can be interpreted not as a cold accident or arbitrary punishment, but as a field of moral development in which conscious beings refine character, responsibility and awareness. The systems language gives this interpretation structure without claiming to settle the mystery beyond dispute.

PART III — SYSTEM ARCHITECTURE TAKEAWAYS

Life, Consciousness, and Re-embodiment

1. Prebiotic Libraries and Migration vs. Origin

Space functions as an active prebiotic factory, delivering raw amino-acid libraries and other chemical building blocks through comets, asteroids, and interstellar material transport. Within this framework, panspermia explains the movement and distribution of life's ingredients across planetary sandboxes, but it does not fully resolve the deeper question of origin. The underlying transition from chemistry to biology remains a fundamental runtime initialization problem.

2. The Humanoid Blueprint and Cumulative Memory

The humanoid form serves as a universal template for unbound moral agency and high-level conscious decision-making. While biological evolution governs local adaptation to environmental conditions, human intelligence introduced a secondary inheritance engine built upon language, education, culture, and external memory systems. Knowledge can therefore accumulate across generations, allowing civilization to advance beyond the limits of purely biological inheritance.

3. Log Closure, Cloning, and Re-embodiment

At the termination of a biological node, the local lifetime record is finalized, evaluated, and archived. The accumulated karmic state is then carried forward through a new initialization template for a subsequent embodiment cycle. In this model, re-embodiment preserves continuity of identity, developmental lessons, unresolved obligations, and long-term state progression across multiple lifetimes.

4. Multi-Tier Evaluation and Non-Physical Governance

Physical existence serves as the initial training and assessment environment for conscious development. Nodes that achieve sustained moral refinement and responsible stewardship may transition into **Tier 2 Non-Physical Governance Domains** for extended evaluation and higher-scope responsibilities. Advancement remains conditional on demonstrated maturity, while misuse of authority may result in recalibration, demotion, or further developmental cycles before additional promotion is permitted.

Part III Core Principle

Life is more than biological chemistry. Biological organisms provide temporary execution environments through which consciousness develops, learns, and accumulates state. Evolution governs physical adaptation, while memory, culture, moral choice, and re-embodiment provide the mechanisms through which conscious development can extend across multiple generations and lifecycles.

PART IV - DIMENSIONS AND THE MEANING OF NEAR

Looking Back:

In **Part III**, we explored how living chemistry transitions into conscious awareness, moral choice, and persistent soul state records. We mapped the multi-tier lifecycle—tracing how a conscious node's choices are logged, finalized, cloned, and evaluated across physical lives and higher non-physical governance tiers.

Moving Forward:

In **Part IV**, we radically expand our understanding of space, distance, and connection. We break free from the low-dimensional trap of viewing "dimensions" merely as physical hallways, redefining a dimension as any independent coordinate needed to describe the complete state of an entity. We explore high-dimensional geometry and network topology—asking how information, influence, and life events travel across deep space via bulk gravity, wormholes, field resonance, and topological entanglement. Part IV reveals a million-dimensional interlinked multiverse where physical distance is merely a shadow, and true proximity is governed by state alignment.

Chapter 16

High-Dimensional Reality and Cosmic Signal Transmission

Note: *Standard quantum mechanics and general relativity describe extraordinarily well-tested aspects of our observable universe. Some additional ideas discussed in this chapter—such as higher-dimensional bulk spaces, traversable wormholes, cross-universe field connections, and cosmic state transmission—remain theoretical or speculative. The Cosmic Architecture uses them as systems analogies for exploring how apparently distant nodes might be related across a deeper structure, not as established mechanisms of multiverse communication.*

When we hear the word "dimension," our minds almost instantly picture physical space. We think of moving left and right, up and down, or forward and backward. Science fiction often exaggerates this concept, depicting a dimension as a glowing portal or a parallel room where alternate versions of us walk around.

However, in physics, mathematics, and our overarching framework of **Cosmic Architecture**, the word *dimension* holds a far broader, more elegant, and deeply practical meaning:

At its core, a dimension is simply a single, independent coordinate required to describe the state or condition of something. Once we stop treating dimensions only as physical directions, the concept becomes much more useful. Within this book's framework, higher-dimensional language helps describe complex states, relationships and meanings that may not fit neatly into ordinary space. This does not prove that reality is a purposeful database; it provides a disciplined way to ask how physical location, history, intention and relationship might coexist in one larger model.

1. What a Dimension Actually Is

To locate a point on a flat sheet of paper, you need two numbers: width and height. This is a two-dimensional grid. To locate a light bulb hanging inside a room, you need three numbers: width, height, and depth. This is our familiar three-dimensional physical space. If you want to schedule a meeting under that exact light bulb, you must provide a fourth coordinate: **time**.

[2D Physical Space] ──▶ Width + Height
[3D Physical Space] ──▶ Width + Height + Depth
[4D Physical Space] ──▶ Width + Height + Depth + Time
[Cosmic State-Space] ──▶ Width + Height + Depth + Time + History + Intent + Karma + ...

However, dimensions do not end with physical movement and time. In mathematics and software architecture, dimensions are used as descriptive variables to define complex systems:

- **A Medical Patient Record:** A doctor's database entry for a patient might track blood pressure, heart rate, body temperature, oxygen levels, cholesterol, and age. In data science, that record represents a **6-dimensional state** describing the patient's health at that exact moment.
- **Modern Artificial Intelligence:** An AI language model does not see words as mere letters. It converts every word or concept into a high-dimensional vector—sometimes using thousands of numeric coordinates—to map its meaning, emotional tone, and relationship to every other word in human history.

When this model talks about millions of "dimensions," it doesn't mean physical hallways you can walk through. Instead, it describes a giant map of possibilities—a **Cosmic State-Space**.

Think of it as a huge map where everything has its own location. Some points show physical details like size or weight. Others represent non-physical qualities like memories, choices, and spiritual traits. This is simply a helpful way to think about reality, not a claim that a real cosmic computer has been found.

2. Geometry vs. Topology: Understanding Nearness

To understand how connections exist across this state-space, we must distinguish between two fundamental branches of mathematics: **Geometry** and **Topology**.

- **Geometry** focuses on physical shapes and measurable distances—such as the exact number of kilometres, miles, or light-years separating two points on a map.
- **Topology** focuses on connections, pathways, and relationships—how elements are linked together regardless of how far apart they sit in physical space.

Consider two people standing on opposite sides of the planet—one in Sydney and one in New York—separated by over 16,000 kilometres of ocean and rock. Geometrically, they are almost as far apart as two humans can be on Earth.

However, if both individuals pick up their smartphones and connect on a video call, their **topological connection** is nearly instantaneous. On a communication network diagram, they are a single hop apart.

[GEOMETRIC DISTANCE] —► Sydney <------- (16,000 Kilometres) -------> New York

[TOPOLOGICAL DISTANCE] —► Sydney <====== (1 Direct Network Hop) ======> New York

3. Signal Transmission: How Information Travels in Our World

Topological nearness sounds simple on paper, but in the real world, a connection cannot exist in a vacuum. Signals require a physical medium or carrier mechanism to travel from Point A to Point B.

To appreciate how information might travel across the cosmos, we must first look at the four primary ways humans send signals across our world:

HUMAN SIGNAL TRANSMISSION MEDIA	
Copper Wires	Electrical pulses moving through physical metal conductors
Fiber Optics	Photons (light beams) bouncing inside glass subsea cables
Electromagnetic	Radio waves & microwaves radiating through the atmosphere
Quantum Optics	Encrypted photon states traversing satellite links

A. Electrical Conduction (Copper Wires)

In early telegraphs and modern landlines, information travels as electrical charges moving through copper wires. Electrons push one another down the metal line, carrying binary signals across continents.

B. Guided Light (Fiber-Optics)

When you load a webpage hosted on a server on another continent, your data does not travel through empty air. It travels as pulses of laser light inside thousands of miles of **Fiber-optic submarine cables** laid along the ocean floor. The light is trapped inside glass filaments, bouncing continuously until it reaches its destination at nearly 200,000 kilometres per second.

C. Radiated Waves (Electromagnetic Fields)

Mobile phones, Wi-Fi, and satellites do not use physical cables. They radiate electromagnetic energy through space. A radio transmitter oscillates electrons, creating ripples in the invisible electromagnetic field that pervades the atmosphere. A receiver antenna miles away detects those field ripples and decodes them back into sound, text, or images.

D. Quantum Satellite Communication

Modern security systems use entangled photons beamed from ground stations to orbiting satellites. Information is encoded directly into the quantum states of light particles, allowing secure key distribution across thousands of kilometres.

In every human technology, **a signal requires an underlying channel or field** through which to propagate.

4. Cosmic Signal Transmission: How Information Travels Across the Multiverse

This brings us to the ultimate question: *If two nodes—two souls, planets, or parallel universes—are topologically linked in the cosmic architecture, what acts as the "cable" or "radio wave" carrying signals across deep space and higher dimensions?*

In the cosmos, where there are no copper wires or glass fibre cables laid between worlds, this book considers several speculative mechanisms that could help us imagine information, influence or state relationship across deep structure. These are not presented as established channels of multiverse communication, but as conceptual bridges between physics, topology and metaphysical possibility:

A. Higher-Dimensional Gravity (The Bulk Highway)

Some theoretical models in physics explore the possibility of additional spatial dimensions beyond the three we experience. In certain brane world models, for example, our observable universe is mathematically represented as a three-dimensional "brane" embedded within a higher-dimensional space often called the "bulk." These models arise from theoretical physics and are not established by current observations. The Cosmic Architecture does not claim to be one of these physical theories. Instead, it borrows the general idea of higher-dimensional geometry as an analogy for exploring how apparently distant regions or nodes might be related within a deeper architecture.

If such higher-dimensional models were ever supported, gravitational effects might become one candidate for thinking about influence beyond ordinary three-dimensional separation. At present, this should be read as a speculative analogy, not as a demonstrated pathway between neighbouring universes.

In this metaphor, gravity acts less like a message sent through ordinary space and more like a change in the shape of the stage itself. The image is useful for imagining cross-layer influence, but it should not be taken as proof that human science has detected a cross-dimensional telegraph.

[UNIVERSE MEMBRANE A]   [UNIVERSE MEMBRANE B]
(3D Physical World)   (3D Physical World)

└ Gravitational Waves Leaking Through "The Bulk" ┘
(High-Dimensional Bridge)

B. Einstein-Rosen Bridges (Cosmic Submarine Cables)

Wormholes, or Einstein-Rosen bridges, are mathematical possibilities within general relativity rather than confirmed cosmic infrastructure. They are useful here as an analogy for shortcut connections: situations where two distant regions could be treated as adjacent within a deeper geometry. If such structures existed in a traversable form, they would challenge ordinary assumptions about distance. For this book, their role is mainly conceptual: they help the reader imagine how topology can matter as much as geometry when thinking about deep connection.

C. Universal Field Resonance (The Shared Substrate)

In modern quantum field theory, particles are not isolated marbles; they are localized ripples on universal fields (like the electron field or electromagnetic field) that stretch across the entire cosmos.

If two parallel universes or entity nodes share the exact same foundational field substrate in higher dimensions, a signal does not need to "travel" from Point A to Point B at all.

Think of two floating corks on opposite sides of a swimming pool. If you tap the water at Cork A, you send a wave through the water that causes Cork B to bob up and down. Cork A didn't send a physical object to Cork B; both corks reacted because they rest on the **exact same underlying liquid medium**.

At the cosmic level, the shared-substrate image suggests how two distant systems might be related through a deeper field-like background. This remains a philosophical extension of field language, not a result already established by quantum field theory.

D. Topological Entanglement Graphs & The Mechanics of Life Events

At the quantum level, entangled systems display correlations that challenge classical intuition while still obeying the limits of usable information transfer. In this book's higher-dimensional metaphor, entangled nodes can be imagined as sharing a graph edge in a deeper state-space, but this is an interpretive model rather than a claim that destiny is physically transmitted by entanglement.

This brings us to a fundamental question: **How does a major life event or sudden shift happen to a human node?**

A major life event can be interpreted through the interaction of two complementary ideas in the model:

1. **The Vector Database Alignment:** The macro-level positioning that places a human node near specific events, environments, and nearest-neighbour nodes.
2. **The Soul Log Database (The Free Will Engine):** A dedicated, continuous logging subsystem that records every choice, action, and moral decision made by a human node in real-time.

Free Will and Incremental Logging

Human beings have been granted **free choice**—the agency to execute actions according to their own will. A human node can choose to act in alignment with its conscience, preachings, and moral integrity, or it can choose to act directly against them.

5. The Soul Log as the Signal State

Earlier chapters introduced the human body as a temporary biological execution and recording node. Its nervous system, brain, sensory organs, and billions of cells continuously respond to the events of physical life. Within the Cosmic Architecture, however, the biological body is not imagined as the permanent store of the conscious node's accumulated history. It is the local 3D runtime through which experience is generated.

The model therefore distinguishes between **biological recording** and the deeper **persistent soul record**. During life, the biological node continuously generates physical, emotional, relational, and behavioural information. Meaningful choices and intentions contribute to the evolving state of the conscious node—the persistent state vector introduced in **Chapter 15.**

This distinction becomes important in a chapter about cosmic signal transmission. A higher-dimensional architecture cannot respond to the changing condition of a conscious node unless that condition is represented in some persistent form. Within the model, the soul record therefore functions as the node's continuing informational state: the changing profile through which choice, consequence, relationship, and future alignment can be interpreted.

This is not a claim that scientists have discovered a literal cosmic database or a detectable communication channel between biological cells and a soul. It is the systems architecture proposed by this book to connect:

Lived Experience → Meaningful Choice → Persistent State Update → High-Dimensional Recalibration

The Non-Obstruction Rule

The model imagines a strict **Non-Obstruction Rule**: conscious agents are allowed to choose freely, while meaningful choices become part of their evolving state.

Accumulated Karma: Significant intentions, decisions, and actions contribute to the node's developing karmic and experiential profile. Karma is therefore represented as accumulated state and consequence rather than as an immediate system of reward and punishment.

Continuous State Updating: During physical life, new experiences and choices continually alter the node's condition. The biological body generates the local experience, while the model imagines its meaningful informational consequences contributing to the persistent soul record.

Cross-Iteration Persistence: When the biological node reaches the end of its lifecycle, the physical body dissolves, but the model imagines the accumulated state being finalized and preserved as part of the continuing conscious-node record.

How a Major Life Recalibration Occurs

As accumulated choices and experiences change the node's state, its wider relational alignment may also change.

Within the model:

1. The node's state vector changes.
2. Its position within the wider high-dimensional architecture changes accordingly.

3. Different relationships, opportunities, challenges, or developmental conditions may consequently become more relevant to that state.
4. Those changes can eventually appear within ordinary 3D life as altered circumstances or major turning points.

The exact geometry behind this idea—how state vectors, similarity, dimensionality, and **nearest neighbours** are represented—is the subject of the next chapter.

6.From Signal Transmission to State-Space Alignment

Once a conscious node's state can change, another question immediately follows: what does that changed state become aligned with? Within the Cosmic Architecture, this is where signal transmission connects to state-space geometry. Changes in a node's accumulated condition may alter which relationships, environments, opportunities, or developmental conditions are most compatible with it.

The word "near" here does not necessarily mean physically close. Two states may be highly related even when the corresponding entities are separated by enormous physical distances. Likewise, two physically adjacent people may occupy very different positions within the model's broader state-space.

Exactly how such high-dimensional proximity works—and how millions of possible characteristics could be represented and compared—is explored in Chapter 17, **Million-Dimensional State-Spaces**.

Chapter 17

Million-Dimensional State-Spaces

Earlier chapters introduced vectors, similarity, nodes, graphs and the idea that two things can be mathematically close even when they are physically far apart. This chapter applies that principle to a much larger question: **how could entire planets, civilizations or universes be represented and distinguished within a multiverse too large for ordinary spatial maps?**

Within *The Cosmic Architecture*, the phrase **million-dimensional state-space** does not mean a million literal physical directions. Instead, a dimension can be understood as one independent property used to describe the state of something. A human being, for example, cannot be described only by physical location. Age, DNA, memories, relationships, education, behaviour, history and countless other factors also help define that person's state. In a sufficiently detailed representation, millions of such coordinates could theoretically be used to distinguish one conscious node from another.

The same principle can be scaled upward. A planet could be described through gravity, atmosphere, oceans, chemistry, biological development, climate, geological history and civilization. An entire universe could require an enormously richer signature containing physical constants, matter distribution, galaxy structures, stellar populations, planetary systems, biological development, causal history and perhaps millions of other parameters. In this sense, a universe becomes conceptually similar to a point within an immense high-dimensional state-space.

This changes the meaning of **near**. Imagine two software engineers, one living in Brisbane and another in London. They may be thousands of kilometres apart physically yet extremely close in a recommendation system if they share similar skills, interests and backgrounds. Meanwhile, the person living next door may be mathematically very different. Physical distance therefore answers one question—*where is it?*—while state-space distance answers another—*how similar is it?*

The same distinction can be applied to worlds. An Earth-like planet thousands of light-years away may be much closer to Earth in state-space than a nearby planet with a poisonous atmosphere, extreme gravity and no complex life. At the universe scale, two universes could likewise be physically inaccessible to one another yet occupy neighbouring regions of state-space because their physical laws, biological development or civilization patterns are highly similar.

This is where the vector-database analogy becomes useful. Modern vector systems can locate relevant records without examining every record sequentially. A query is converted into a mathematical representation, and the system identifies nearby or similar states. *The Cosmic Architecture* extends this idea speculatively. Instead of asking, "Which document is most similar to this document?" a higher-level architecture could conceptually ask, "Which planetary systems most closely match this biological condition?" or "Which civilizations are approaching this developmental threshold?"

The million-dimensional model therefore provides a way of thinking about **cosmic addressability**. A universe, sector, planet or conscious node could theoretically possess a unique multidimensional signature. Higher tiers would not need to search every world one by one; they could conceptually identify relevant states through their position within the larger state-space.

This does not mean similarity and connection are the same thing. **Chapter 4** distinguished vectors from graph relationships: vectors describe how similar two nodes are, whereas graph edges describe whether those nodes are actually connected. Two universes could therefore be very similar without sharing a direct relationship, while two very different universes might still be linked through branching, inheritance or a common higher-level structure.

The distinction becomes useful when considering traditional descriptions of higher or lower realms. Human language tends to interpret such terms spatially, as though a higher realm must exist physically above Earth. Within the state-space model, however, "higher," "lower," "near" and "far" need not always refer to ordinary physical distance. They may instead describe different states, levels of access, developmental conditions or positions within a broader architecture. Just as multiple television channels can reach the same receiver without being stacked physically above one another, different domains could theoretically coexist within a larger system while remaining inaccessible through ordinary human perception.

The model also helps clarify the idea of a multiverse. Popular illustrations often show universes as bubbles floating beside one another, which encourages us to imagine ordinary distance between them. *The Cosmic Architecture* instead asks us to imagine a vast landscape of states. Universes with similar properties would cluster conceptually, while radically different universes would occupy distant regions of that state-space. A universe does not therefore need to be physically beside another universe to be its nearest neighbour.

This becomes especially important when the book later discusses higher-tier governance. The million-dimensional state-space provides the **addressing layer**. It helps answer: *Which node, sector or universe state is relevant?* It does not by itself explain how information travels to that target. That is a separate architectural problem. Within the broader framework, the distinction can be summarized simply: **state-space identifies the destination, graph relationships describe actual connections, and the deeper cosmic substrate provides the proposed shared layer through which higher-tier access might occur.**

This separation is important because it prevents the vector analogy from being overstretched. A vector database can explain how a complex state might be represented and queried; it does not prove that universes communicate through vectors or that reality itself is literally a database. The analogy is a conceptual tool.

The larger proposal is therefore straightforward. If reality contains an enormous number of planets, civilizations, conscious nodes or universe domains, physical coordinates alone may be insufficient to describe how they relate. A richer state-space could provide another form of geometry—one based on similarity, condition and structure rather than location alone.

High-dimensional mathematics, vectors and nearest-neighbour searches are real computational concepts. What remains speculative is the extension of those ideas to consciousness, universes and cosmic governance. *The Cosmic Architecture* uses them not as proof, but as a way of making the hypothesis more precise.

The central idea of this chapter can therefore be stated simply:

Physical space tells us where something is.
State-space tells us what kind of state it is.
Graphs tell us what it is connected to.
A deeper substrate, if such a layer exists, would determine how different domains could interact.

That is the meaning of the **million-dimensional state-space** within *The Cosmic Architecture.*

The million-dimensional state-space also provides a conceptual query model. If every universe, planetary sector, civilization, or conscious node possesses a sufficiently rich multidimensional signature, a higher-level architecture would not need to search the multiverse sequentially. It could instead identify states matching particular conditions—much as a vector search locates relevant neighbours within a high-dimensional database. A query might conceptually ask for Earth-like environments, civilizations approaching a developmental threshold, or nodes matching a particular state. The analogy does not imply that the cosmos literally operates a software database; it provides a way of imagining how an enormously complex architecture could identify relevant states efficiently.

This query model explains identification, not communication. State-space helps determine *which* node or universe is relevant. The deeper substrate discussed later in the book provides the separate speculative model for how higher tiers might access or communicate across universe domains.

PART V - MIND, MEMORY AND INFORMATION

Looking Back:

In **Part IV**, we expanded our geometry into high-dimensional state-space. We learned that nearness in the *Cosmic Architecture* can be represented through state similarity and topological graph connections rather than physical distance alone. We also explored speculative possibilities—including higher-dimensional gravity, deeper field relationships and shared substrates—for imagining how apparently distant sectors might participate in a larger architecture.

Moving Forward:

Part V turns its attention to the underlying storage and processing substrate of the cosmos. We draw a fundamental distinction between Chemical Architecture (the temporary biological hardware) and Information Architecture (the enduring software protocol). We explore how the universe eliminates external data centres by encrypting history directly into subatomic geometry, DNA, and consciousness vectors—operating as a self-querying, living ledger. Finally, we examine the processing ceilings of 3D brain wetware and how gifted minds tune into over-the-air knowledge downlinks to drive civilizational breakthroughs.

Chapter 18

Information Architecture vs. Chemical Architecture

When we observe the natural world, our eyes immediately register physical matter. We see oceans, mountains, trees, animals, and human bodies. Chemistry describes the elements that compose these structures: hydrogen, oxygen, carbon, nitrogen, calcium, and iron. Physics describes the forces acting upon them: gravity, electromagnetism, and nuclear interactions.

Because matter is visible and tangible, human thinking naturally treats physical chemistry as the ultimate baseline of reality. We assume that if something exists, it must be made of atoms.

1. Chemical Architecture (The Physical Substrate / Hardware)

Serves as the temporary local runtime environment:

- **In Plain English:** Think of a "local runtime environment" like a physical stage or a gaming console sitting in your living room. The console provides the physical plastic, circuits, memory chips, and electricity needed for a video game to run.
- **In the Cosmic Architecture:** Our 3D physical world—made of carbon chains, liquid water, planetary atmospheres, and DNA molecules—is simply the temporary physical "stage". It is the local hardware sandbox where living beings move, interact, experience physical limits, and execute their daily choices. The hardware decays over time, but it provides the active physical environment needed for life to play out.

2. Information Architecture (The System Logic / Software Protocol)

The System Logic / Software Protocol:

- **In Plain English:** Imagine a web browser or a mobile app. The physical glass screen and battery in your hands are the *hardware*, but the rules governing how the app works—the coding, the math, the algorithms, and the network rules—are the *software protocol*. You cannot touch a software rule with your fingers, yet it completely dictates what happens on the screen.
- **In the Cosmic Architecture:** Long before physical atoms ever assembled into the first living cell, the underlying rules of nature were already written into the fabric of reality. The **Information Architecture** consists of these unchangeable master rules: the exact strength of gravity, the mathematical symmetries of chemistry, the speed of light, and the high-dimensional logs that record human choices and moral history. Chemistry does not invent life out of nowhere; it simply executes the invisible software protocol configured by the cosmos.

Comparison at a Glance

Component	Technical Term	Everyday Analogy	Cosmic Architecture Role
Chemical Layer	**Local Runtime Environment**	The Physical Computer / Game Console	The temporary 3D biological body and physical world where choices occur.
Information Layer	**System Logic / Software Protocol**	The Operating System & Game Code	The underlying mathematical laws, physical constants, and state-logging rules.

To build a rigorous *Cosmic Architecture*, we must separate the **Chemical Architecture** from the **Information Architecture**.

The Chemical Substrate: The Local Runtime Engine

The Chemical Architecture represents the physical medium through which life and planetary environments operate.

Consider liquid water. As explored in Chapter 39, Earth became a water world because possesses specific thermodynamic properties: a high specific heat capacity, a strong dipole moment, and the ability to act as a universal polar solvent. Water creates a buoyant 3D fluid matrix where prebiotic molecules can collide, dissolve, and self-assemble without collapsing under gravity.

Consider DNA. Biological life on Earth uses double-helical nucleic acid chains built from adenine, thymine, cytosine, and guanine. This molecular arrangement allows genetic information to be copied, transcribed into RNA, and translated into functional protein structures.

Water, carbon chains, and DNA are brilliant engineering solutions. But they are **chemical hardware**.

[Software / Protocol] DNA Code / Regulatory Switches

[Physical Hardware] Amino Acids / Proteins / Water Substrate

A software engineer looking at a web application does not confuse the Python or Go code with the copper wiring inside the server rack. The copper wire carries electrical current, but the logic resides in the software protocol.

Similarly, DNA is not the ultimate source of life's architecture; it is the **chemical storage medium** through which biological logic is executed in our 3D physical sandbox.

The Information Architecture: The Rules Behind the Substrate

What determines *why* chemistry behaves with such precision? Why do carbon atoms form stable organic rings? Why do hydrogen and oxygen combine to create a polar molecule with a high dipole moment? Why do developmental switches in the human genome allow the exact same DNA to construct both a retinal cell in the eye and a cardiac muscle cell in the heart?

The answer belongs to the **Information Architecture**.

The Information Architecture consists of the underlying rules, mathematical symmetries, physical constants, and high-dimensional state-space protocols that dictate how matter behaves.

1. **The Parameter Baseline:** Physical constants—such as the speed of light (c), the gravitational constant (G), Planck's constant (h), and the fine-structure constant (α)—are not physical objects. They are **systemic configuration values** that establish the baseline operational limits of our universe.
2. **The Possibility Space:** Long before the first biological cell emerged on Earth, the rules of atomic bonding, thermodynamics, and self-organization already existed in the mathematical structure of reality. The Chemical Architecture did not invent life; it discovered and executed a pathway made possible by the Information Architecture.
3. **The Semantic State:** As established in our Part I primitives, physical matter decays through entropy. Stars burn out, biological bodies age, and chemical structures dissolve. But the high-dimensional state log—the moral choices, intentional trajectories, and accumulated wisdom generated by a conscious Node—is recorded within the Information Architecture.

HARDWARE vs. SOFTWARE CONSERVATION

PHYSICAL DECAY (Chemical Layer):

- *Biological Body Dissolves into Soil / Atoms (Entropy)*

STATE CONSERVATION (Information Layer):

- *Soul State Log Preserved in State-Space (Conservation)*

When a human life ends, the chemical hardware returns to the planetary ecosystem. The Information Architecture ensures that the high-dimensional state vector is preserved, cloned, or evaluated for the next operational cycle.

By separating Chemical Hardware from Information Logic, we escape the primitive materialist assumption that reality ends when physical matter breaks down. Chemistry is simply the local, temporary stage; Information is the enduring architecture.

Biological Biology as Edge Computing

In modern enterprise software, engineers deploy **Edge Computing** (local devices like smartphones or IoT sensors) to process data on the ground without overloading the central data centre.

In *The Cosmic Architecture,* **biological bodies operate as the Edge Execution Nodes of the multiverse:**

- **Hardware (The Body):** Local carbon-based terminal running on daily chemical calories.
- **User Interface (Sensors & Telemetry):** Pain acts as localized hardware alerts; empathy functions as network bandwidth protocols.
- **Local Runtime (Free Will):** Operates autonomously in 3D space under local rules, passing finalized summary telemetry (soul logs) back up to the central archive upon node termination.

Viewing biology through edge computing clarifies why physical bodies are temporary. The local biological hardware inevitably degrades over time due to entropy, but the data generated during its operational lifecycle—its character, moral choices, and state vector—is preserved and uploaded to the higher-tier cosmic network.

Chapter 19

Cosmic Memory – The Atomic Database, Real-Time Query Engine, and Living Ledger

Whenever we think about information, we almost automatically picture human technology. We visualize silicon chips, magnetic hard drives, cloud data centres, and global server farms containing millions of flashing racks. Every technology humanity has ever built follows the same basic assumption: if information is valuable, it must be saved inside a separate physical container.

We build databases, generate backups, and implement disaster recovery protocols because, in our daily experience, information can easily be lost, corrupted, or erased. To us, storage seems so fundamental that we rarely stop to question whether it is necessary for reality itself.

Perhaps that assumption tells us more about our current stage of technological infancy than it does about the universe.

1. The Evolutionary Ladder of Storage

Every human era develops tools appropriate for its own level of understanding. Thousands of years ago, ancient civilizations carved stories and laws into stone slabs because stone survived longer than human memory. Later, we invented papyrus, parchment, and paper. Libraries allowed knowledge to bridge generations, and the printing press multiplied books by the millions.

In the late twentieth century, digital technology transformed paper into electronic bits. Cloud computing distributed those bits across global server networks, and today, quantum computing is exploring entirely new ways of representing information through subatomic states.

[Stone Slabs] ──► [Paper & Libraries] ──► [Silicon & Cloud] ──► [Quantum Qubits] ──► [Native Atomic Memory]
(Primitive Era) (Cosmic Architecture)

Each generation naturally believes it has discovered the ultimate, most advanced method for preserving knowledge. Yet history repeatedly proves otherwise. If human civilization survives for another million years, our descendants will look back at today's cloud data centres in the exact same way we view ancient clay tablets.

Now, step back and imagine a cosmic system that has evolved not for millions, but for trillions of years across multiple cosmic cycles.

Would such a system still build physical computers? Would it still rely on external databases or silicon servers? Would it still process information using bits, bytes, or qubits?

Within the Cosmic Architecture, the answer need not be yes. The model instead imagines a stage of development in which information no longer needs to be stored in a separate external container, because physical state and information become deeply integrated.

2. Atoms as Rows: The Native Atomic Database

Human technology creates a sharp distinction between the container and the content. A paper book is a container; the written words are the information. A magnetic hard drive is a container; the software code is the information.

Because of this separation, we look at the cosmos through a primitive lens. We assume a star is merely a ball of plasma, a planet is just rock and metal, and an atom is simply a mechanical marble.

In the framework of *The Cosmic Architecture*, this artificial divide dissolves. The universe does not need to record events onto an external hard drive because atoms and physical energy states **are** the database. Over trillions of years and across successive cosmic deployments, the universe perfected an unimaginable level of native encryption and data partitioning.

Atoms as Structured Records

To picture how the universe stores information natively, imagine a simple digital spreadsheet or filing system:

- **The Database Table (The Element Group):** Think of a category or spreadsheet tab—like the "Iron" list or the "Carbon" list.
- **The Row (A Single Atom):** Each individual atom is like a single row on that spreadsheet, representing one specific entity.
- **The Primary Key (Atomic Number / Unique ID):** In computing, a "primary key" is a unique ID tag that tells the system exactly what an item is. For an atom, this is its number of protons. If it has 6 protons, it is permanently identified as Carbon; if it has 26, it is Iron.
- **The Columns (Properties and History):** In a spreadsheet, columns hold specific details about a row (such as age, weight, or balance). In an atom, these "columns" record its subatomic details:
    - *The Neutron Count (Isotope):* Acts like a date-stamp recording where and when the atom was forged or exposed.
    - *The Electron Shells:* Act like an energy tracker, showing how much heat or light the atom has recently absorbed.
    - *Quantum Spin:* Holds ultra-fine, encrypted state details.

Molecules as Connected Records (Relational Joins)

In computer science, a "relational join" simply means linking two different rows together to create a richer profile. When one Oxygen atom bonds with two Hydrogen atoms to form water (H_2O), the cosmos is executing a chemical join: it links individual records together into a new, functional molecule capable of storing and transferring biological and thermodynamic information.

3. Executing the Query: How Humans and the Cosmos Read the Database

If the physical universe is an encrypted, atomic database, how is that data retrieved? In human software engineering, an application developer writes a query (like SQL) to fetch specific information from a relational table:

SQL

```
SELECT user_id, experience_points, clearance_level
FROM user_state_table
WHERE status = 'Active';
```

A human database query is a structured request sent to a computer to fetch, filter, and compute specific information. But querying the cosmic database happens in two distinct ways: how *human science* reads it from the outside, and how the *cosmos* executes it natively from the inside.

A. How Human Science Queries the Cosmos (Reading the Past)
We do not need a keyboard to read the history of an atom; we use physics and chemistry. Scientists routinely query the atomic database using proven analytical methods to extract billions of years of history:

- **Radiometric Dating (Reading the Historical Timestamp):** Certain unstable atoms (like Carbon-14 or Uranium) decay into stable atoms at an exact, unchangeable mathematical rate. By measuring the ratio of decayed atoms to original atoms in a rock or fossil, geologists are essentially reading the atom's internal time-log. This allows us to determine the exact age of an ancient artifact, a meteorite, or the Earth itself.
- **Spectroscopy (Reading the Light Fingerprint from Trillions of Miles Away):**

  A telescope is not a microscope, and physical atoms from distant stars do not fly across space into our lenses. So how can astronomers know the exact chemistry of a star billions of trillions of kilometres away?

  The secret is that light carries the receipt.

  When an electron inside an atom absorbs or emits energy, it only interacts with very specific colours (wavelengths) of light. Because every chemical element has a completely unique electron structure, every element leaves a distinct pattern of dark or bright lines across the spectrum—like a unique barcode.

When starlight travels across the void and enters a telescope, astronomers pass that light through a prism (a spectrograph). They are not looking at the physical atom; they are reading the barcode stamped into the light itself. Just as a supermarket scanner reads a product's barcode without opening the box, a telescope reads the light to know the star's exact chemical elements, temperature, pressure, and velocity.

B. How the Universe Queries Itself (Running the Present)

A **Cosmic Query**, while far beyond human visual imagination, operates on a fundamentally similar architectural principle to our SQL code: it is an automated information retrieval process that evaluates physical, biological, and conscious states in real time.

You do not need a terminal to execute a cosmic query. The query engine is embedded directly into the fabric of physical and metaphysical laws. How do trillions of unthinking atoms manage to keep a living person breathing, walking, and thinking? They do it through continuous, microscopic interactions that function exactly like software queries and system dependencies:

- **Biochemical Micro-Queries (Gene Regulation):** When a biological cell reads a strand of DNA to manufacture a protein, it executes a localized micro-query. Enzymes scan base pairs, match complementary nucleotides, and verify conditions before initiating transcription. The cell is querying the biological table: *"Is this specific sequence of code needed right now?"*
- **Cellular Signalling as Network Messages (Dependency Checks):** When your blood sugar drops, pancreas cells release insulin molecules. These molecules travel through the bloodstream and bind to receptors on liver and muscle cells. This interaction is a network dependency check: *"Does the body need energy storage updated?"*

- **Autonomous Coordination:** No single central brain tells every individual atom what to do. The atoms interact locally according to physical laws, yet their collective, interconnected execution maintains a living, breathing human entity.
- **Karmic Queries (Real-Time Logging):** When a human node exercises free will—choosing compassion over cruelty—the action shifts their internal state. The architecture executes an immediate state query that logs the net moral value into their Soul Record, dynamically updating their trajectory across the system.

4. How Matter Remembers Where It Has Been

Consider an atom of iron currently circulating in a red blood cell through your body:

- **Forged in Cosmic Furnaces:** Billions of years ago, it was synthesized inside the superheated core of a massive star and scattered across space in a supernova.
- **Gathered into Worlds:** Gravity drew that debris together along cosmic dust lanes to form our solar system, the Earth, and its crust.
- **Woven into Biology:** Soil minerals were drawn into plant life, cycled through the food web, and assembled into your physical anatomy.

That iron atom carries its origin tag, energy history, and chemical bonds within its physical structure—an unalterable historical log of everywhere it has been.

4. The Body as a Real-Time Recording Terminal

A sceptical reader or physicist might naturally ask: *How could a human body possibly log every millisecond of action, thought, and intent throughout an eighty-year lifetime without running out of physical storage space?*

The answer lies in recognizing that matter itself is the ultimate storage medium.

The human body does not require an external camera or physical hard drive to record its journey. The recording is built directly into the physical hardware:

- **Biological Telemetry (Brains & Epigenetics):** Experiences and learning physically reshape neural connections, while epigenetic mechanisms can alter patterns of gene regulation in response to developmental and environmental conditions. In this ordinary biological sense, the body's present state carries traces of its history.
- **Atomic State Information:** Atoms and molecules possess measurable physical states—such as energy levels, spin states and chemical configurations—that change through interaction. These states contain information about a system's present condition and may preserve traces or correlations from earlier interactions. This does not mean that every atom contains a readable recording of everything that has ever happened to it; the Cosmic Architecture extends this scientific fact into a speculative analogy of deeper cosmic memory.

During physical life, the body operates as an autonomous **Edge Execution Node**—logging real-time micro-events locally within its physical wetware.

When the biological body reaches its limit and dies (node termination), entropy breaks down the temporary chemical hardware. However, the central architecture executes an automated log closure: the net moral telemetry, character traits, and state updates generated during that lifetime are extracted, compressed into a high-dimensional **Master Soul Vector**, and within the model committed permanently into the foundational fabric of spacetime under zero-trust security protocols.

Physical matter is merely the temporary recording tape; the compressed state vector is the permanent archival record.

6. Why a Newborn Already Knows How to Live (Embodied Instinct)

This proposed network of atomic and genetic memory provides the answer to one of life's most beautiful mysteries: *Why does a newborn human baby already know how to survive without being taught?*

A newborn infant arriving in the world does not need to attend a lecture on human anatomy or take a training course:

- It instantly knows how to blink, focus its eyes, and move them toward light.
- It knows how to coordinate its respiratory muscles to take its first breath.
- It executes the complex, multi-muscle motor reflex of breastfeeding the moment it touches its mother.

BILLIONS OF YEARS OF EVOLUTION (Trial, Error, Adaptation)

▼

ENCODED INTO CHEMICAL SEQUENCES (DNA & Molecular Geometry)

▼

BUILT INTO BIOLOGICAL HARDWARE (Brainstem & Reflex Circuits)

▼

INSTANTANEOUS HUMAN INSTINCT (Breathing, Sucking, Eye Movement)

Where did that knowledge come from? It did not magically appear from thin air.

That infant is executing instructions written, tested, refined, and preserved over billions of years of biological evolution. Every ancestor that successfully survived, adapted, and reproduced passed forward chemical blueprints—encoded in DNA—that compress those ancient lessons into structural reflexes.

The newborn's body is querying an ancient, inherited repository stored directly inside its physical chemistry. The atoms in its body are following rules that took cosmic epochs to write.

7. Human Engineering versus Cosmic Architecture

To understand how profound this is, compare how humans store knowledge with how the universe does it:

Feature	**Human Method**	**Cosmic Architecture Method**
Storage Medium	External (Paper, Hard Drives, Clouds)	Internal (Atoms, DNA, Physics, Spacetime)
Separation	Object and Data are separate	Object and Data are identical
Maintenance	Requires constant power, repairs, and updates	Preserved directly through natural law and structural state
Execution	Needs an external user or computer to read it	Self-executing; matter acts on its own internal state

Human civilization builds separate places to store data because our technology is still primitive. But an advanced Cosmic Architecture has no need for a giant central server room floating in space.

The Core Insight: The universe does not need a separate database because reality itself is the storage system. The stars carry the history of chemistry. Gravity organizes relationships. DNA carries the history of biology. Consciousness carries the history of experience.

8. Telemetry for the Higher Governance Tiers

Why are these cosmic queries executed continuously?

In cloud enterprise architecture, low-level microservices aggregate operational data and pass summary reports up to high-level management dashboards so system administrators can monitor platform health, detect anomalies, and allocate compute resources.

[Lower Atomic Tables] ──► [Automated Cosmic Queries] ──► [Aggregated Telemetry] ──► [Higher Administrative Tier]
(Atoms, DNA, Choices) (Real-Time Data Extraction) (Net System Reports) (Devas / System Managers)

In *Cosmic Architecture*, cosmic queries aggregate data from lower physical and biological tables and feed that information up to the Higher Administrative Tier (what ancient traditions call Devas, Lokapalas, or System Managers):

- **No Manual Micromanagement:** Higher administrators do not need to manually inspect every single atom or micro-decision. They read aggregated query outputs—the net state logs of conscious nodes, biospheres, and physical sectors.
- **Evaluating Evolution and Readiness:** Just as a cloud platform uses telemetry queries to determine when a server cluster needs an upgrade or a restart, the upper hierarchy uses cosmic query outputs to evaluate spiritual readiness, govern karmic re-embodiment, or determine when a local planetary project phase (like Earth) requires recalibration.
- **Eliminating Waste:** Because cosmic queries continuously digest and pass information upward, nothing experienced in physical reality is ever lost or wasted. Every struggle, moral breakthrough, emotional experience, and biological innovation is logged, queried, and preserved by the broader system.

9. The Present Already Contains the Past: Natural Physical Archives

Before we ask whether reality maintains an abstract or high-dimensional record, we must first recognize that in an ordinary scientific sense, the physical universe already remembers. Memory is not magic; it always requires a carrier, a structure that preserves information, and a process through which the past affects the future.

A diary remembers because ink marks remain on paper. A computer remembers because physical energy states are held inside magnetic or silicon media. A brain remembers because experiences reshape neural connections.

Nature uses this exact same principle at every scale:

- **Earth's Geological Memory:** Fossils preserve evidence of ancient life; geological strata record volcanic eruptions and climate shifts; ice cores trap samples of primordial atmospheres; and tree rings log centuries of changing environmental conditions.
- **Biological Memory:** DNA acts as an evolutionary archive, holding inherited code modifications passed down across millions of generations.
- **Astronomical Memory:** The night sky is a living historical archive. Because light travels at a finite speed (c), looking at distant stars means receiving signals emitted millions or billions of years ago. The sky naturally contains multiple epochs of cosmic history simultaneously.
- **Chemical Memory:** The iron atom inside a human red blood cell carries a history extending back to stellar fusion inside massive stars that exploded long before Earth existed.

In this ordinary, observable sense, the universe does not need an external hard drive—its current physical state IS the unalterable record of its history.

[NATIVE COSMIC INFORMATION LAYERS]
|
├── **1. PHYSICAL LAYER** ──▶ Quantum States, Correlations & Physical Structure
├── **2. BIOLOGICAL LAYER** ──▶ DNA, Epigenetics & Molecular Signatures
└── **3. METAPHYSICAL LAYER** ─▶ Consciousness State Vectors (Karmic & Soul Records)

These three layers summarize the chapter's central idea. At the physical and biological levels, nature demonstrably preserves information through present states, structures, correlations, DNA, neural organization, geological records and other physical carriers. The Cosmic Architecture then takes one further speculative step: it asks whether conscious experience and moral history might also persist within a deeper informational layer of reality. The first two layers are grounded in observable natural processes; the third remains a philosophical hypothesis of this framework.

Chapter 20

Biological Wetware Ceilings, Gifted Minds, and Over-the-Air Knowledge Infusions

The human brain is one of the most remarkable biological structures in the known universe. Weighing roughly 1.5 kilograms—a tiny fraction of total body weight—it consumes nearly 20% of the body's entire daily energy budget. Every building, bridge, library, scientific theory, digital application, and spacecraft capable of leaving Earth began as a dynamic pattern of neural activity inside a human brain.

For thousands of years, human philosophy assumed that knowledge was gained solely through slow, linear experience: we observe the physical world, make trial-and-error mistakes, record what works, and gradually accumulate wisdom across generations. Modern neuroscience expands this view by showing how the brain continually reshapes its neural connections in response to new learning.

Yet, when we look at the breathtaking leaps in human history, a deeper question emerges: **How do revolutionary discoveries occur?** *Is human invention merely the result of local biological effort, or is human technology a tiny, early glimpse of a far grander cosmic architecture?*

1. Human Technology as Tiniest Replicas of Cosmic Science

As software developers, data engineers, physicists, and scientists, we take immense pride in our modern achievements. We build multi-tenant cloud platforms, write intricate mathematical proofs, manage Git version control repositories and design automated CI/CD deployment pipelines

However, the Cosmic Architecture invites us to hold these accomplishments with profound intellectual humility.

[COSMIC ARCHITECTURE]

(Manipulates spacetime metrics, gravity, field resonance, and subatomic memory natively)

|
▼ MICRO-SCALE HUMAN REPLICATION
|

[HUMAN TECHNOLOGY]
(Computers, Git Repositories, CI/CD Pipelines, Microprocessors, Written Mathematics)

Maybe our current human inventions—our digital computers, software repositories, CI/CD release pipelines, engineering machines, and scientific disciplines—are the **tiniest of the tiniest replicas of unimaginable cosmic science and mechanisms operating at the cosmic level**.

Mechanisms Beyond Physical Hardware

When we speak of "cosmic infrastructure" or "deployment pipelines in this proposed architecture," it is critical not to fall into the naive trap of imagining giant steel server racks or mechanical assembly lines floating in deep space. The cosmos has no need for physical keyboards, silicon chips, or copper cables.

The universe operates through native, high-dimensional mechanisms that achieve at a cosmic scale what our primitive machines attempt at a micro scale:

- **Gravity as the Primary Instrument**: Rather than using mechanical arms or physical press machines, the cosmos uses **gravity and space-metric curvature** to compress raw gas clouds, forge heavy elements, shape solar systems, and steer planetary orbits.
- **Direct Matter Coupling**: Higher-tier governance structures do not use physical bodies or manual tools. They interact directly with the fundamental field substrate—coupling with matter at will, guiding density fluctuations to seed galaxies, and fine-tuning chemical affinities so self-organizing life can emerge.

- **Systemic Ecosystem Steering**: The cosmos utilizes fundamental forces to execute systemic shifts—shaping planetary climates, steering cometary delivery vectors, and triggering localized environmental recalibrations to guide ecological and human evolution.

What we call "computer science," "physics," or "biology" are simply human discovery frameworks—our early, localized attempts to map the master software protocol and engineering rules already governing reality.

2. Artificial Intelligence: The Synthetic Extension of the Edge Node

Today, artificial intelligence marks the next leap in this evolutionary ladder.

AI is not merely a static digital filing cabinet. It acts as an active **semantic synthesizer**—a tool that reads, indexes, and finds hidden patterns across billions of books, medical research papers, and mathematical proofs. It maps ideas across multi-dimensional concept spaces, spotting connections across different fields in seconds that would take a human researcher lifetimes to calculate.

This is the externalization of pattern recognition. Just as telescopes extended our limited eyesight and cranes multiplied our physical muscle, artificial intelligence serves as an edge-compute amplifier for human thought.

Crucially, AI carries no conscious awareness, moral agency, or soul record. It remains hardware and tooling. By lifting the burden of routine data processing and calculation off biological wetware, it frees human conscious nodes to focus on the work machines cannot perform such as ethical discernment, authentic empathy, creative inquiry, and responsible moral stewardship of our planetary sandbox.

3. The Biological Wetware Ceiling

Every physical instrument has operational boundaries. A telescope detects only the signals its hardware can receive. A computer processes information according to the capabilities of its processor and memory. The human brain, extraordinary as it is, is also biological hardware operating within physical constraints.

Our senses reveal only a small portion of physical reality. Human eyes, for example, cannot see radio waves, infrared radiation, ultraviolet light, X-rays, or gravitational waves, even though instruments can detect many of them. Similarly, our brains evolved to navigate ordinary three-dimensional environments. Mathematics allows us to describe structures and state spaces far beyond what we can directly visualize.

The brain nevertheless overcomes many of its individual limitations through **cumulative knowledge**. Language, writing, education, mathematics, scientific collaboration, and technology allow one generation to begin where previous generations ended.

There is also a powerful internal mechanism for sudden insight. A person may consciously struggle with a problem, leave it unresolved, and later experience an unexpected solution. Psychology explains much of this through memory, pattern recognition, association, and unconscious processing. The brain may continue reorganizing information even when we are no longer consciously thinking about the problem.

But the *Cosmic Architecture* asks an additional question:

Is all meaningful information available to consciousness necessarily generated or stored inside the local biological brain?

If consciousness is part of a wider informational architecture, as this book proposes, the brain might be more than a self-contained processor. It could also function as a **biological interface**—normally processing local sensory information, but under exceptional conditions capable of coupling with information carried through deeper layers of reality.

This leads to the concept of the **Prepared Mind**.

A person who has spent twenty years studying mathematics possesses a neural framework capable of interpreting mathematical relationships that would be meaningless to someone without that preparation. A possible cosmic information transfer therefore need not download an entire subject into an empty brain. It might instead provide the **missing relationship, pattern, or connection** that allows years of accumulated knowledge suddenly to reorganize into a breakthrough.

The model therefore does not replace human learning.

Preparation builds the receiver. Insight completes the pattern.

4. Over-the-Air Knowledge Infusions and the Prepared Mind

Human history contains remarkable examples of discoveries that appeared as sudden flashes after long periods of preparation. Scientists, mathematicians, inventors, artists, and spiritual thinkers have sometimes described solutions or insights as arriving unexpectedly rather than through an obvious conscious sequence of reasoning.

The conventional explanation—unconscious processing and accumulated expertise—remains entirely valid.

The *Cosmic Architecture* asks whether some exceptional cases could contain an additional layer.

Within the model, a highly prepared brain can be imagined as a **Catalyst Node**. When civilization reaches particular developmental thresholds, information already present within the wider Cosmic Architecture could become accessible to individuals capable of translating it into mathematics, language, technology, art, or ethical insight.

This is analogous to an **over-the-air update**, but the analogy should not be misunderstood. The recipient does not necessarily receive complete sentences or equations like an email attachment. The transfer might instead consist of an informational pattern—a relationship that causes previously disconnected knowledge inside the prepared brain to become suddenly coherent.

This gives us a more interesting physical question:

Could Gravity Carry Information?

Gravity deserves particular attention because it is fundamentally different from simply throwing matter from one location to another.

In Einstein's General Relativity, gravity is associated with the geometry of spacetime itself. Accelerating massive objects can produce **gravitational waves—disturbances in spacetime that propagate outward from their source**.

These waves genuinely carry information.

When gravitational-wave observatories detect signals produced by merging black holes or neutron stars, scientists can analyse the signal and infer properties of systems located enormous distances away. The shape, frequency, amplitude, and evolution of the waveform contain information about the physical event that produced it.

This establishes an important principle:

spacetime disturbances can carry physically meaningful information about their source.

The *Cosmic Architecture* takes that established principle one speculative step further.

If a sufficiently advanced cosmic intelligence could deliberately manipulate extremely subtle properties of spacetime or gravitational fields, could those variations be **modulated** to encode information?

Human technology already uses the same general principle with other physical carriers. Radio communication does not send words physically flying through space. We deliberately vary—or modulate—an electromagnetic wave so that its changing pattern represents information. A receiver detects the pattern and reconstructs the message.

The speculative analogy would therefore be:

Information → Encoded Pattern → Spacetime/Field Modulation → Compatible Receiver → Decoded Information

The proposal is not that naturally occurring gravitational waves from black holes secretly contain human language. Rather, the Cosmic Architecture asks whether an intelligence capable of manipulating the deeper geometry or fields of reality could use them as a **native communication layer**.

This is conceptually similar to the science-fiction idea portrayed in *Interstellar*, where gravitational effects become a means of encoding information across otherwise inaccessible boundaries. The film does not demonstrate that such communication is physically possible, but the underlying thought experiment is useful: **if gravity can influence a region that another ordinary signal cannot reach, could controlled variations in that influence become an information channel?**

Within the *Cosmic Architecture*, the answer is proposed as a possibility.

Why Not Simply Send Matter?

This also explains why a cosmic knowledge transfer would not require physical particles carrying instructions to fly across the universe and enter someone's brain.

The brain is already embedded within physical reality. Its neurons, atoms, electromagnetic activity, and surrounding gravitational environment all exist within the same underlying spacetime and field structure.

If information could be encoded into a deeper physical substrate, the receiving brain would not necessarily require new matter. It would need only to **respond differently to an incoming pattern**.

Think of a radio.

A radio station does not deliver new components into the receiver every time music plays. The hardware is already present. An incoming signal changes the state of that hardware, and the receiver converts the pattern into sound.

The Cosmic Architecture proposes a similar conceptual relationship:

the biological brain provides the local hardware; the deeper substrate provides the potential communication channel.

But Would Mining Matter Damage the Cosmic Information Layer?

This raises another important architectural question. If cosmic information is distributed through the fabric of reality, what happens when humans extract silicon, iron, lithium, stone, or other materials? Are we consuming parts of the cosmic database?

The model distinguishes **matter from the substrate in which matter exists.**

The Ocean Analogy (Best for the Quantum Field Theme)

To understand the difference between matter and the substrate, look at the ocean.

- **The Substrate** is the vast, deep, continuous body of water.
- **The Matter** represents the visible waves, whirlpools, and sea foam crashing on the surface.

When a wave crashes against the shore and dissolves, the wave loses its shape, but the ocean itself is not damaged or depleted. Mining silicon, building a skyscraper, or burning a piece of wood just rearranges the "waves" of physical reality. The deep cosmic ocean beneath it—the spacetime substrate—remains perfectly intact and unbroken.

Mining silicon does not remove a piece of spacetime. It rearranges matter already located within spacetime. The silicon may become a computer chip, later be recycled, melted, chemically altered, or dispersed, while the underlying spacetime continues to exist.

The Cosmic Architecture therefore should not picture its deepest information store as a conventional database in which:

Atom A = File 1
Atom B = File 2
Rock C = File 3

Such an architecture would be extraordinarily fragile.

Instead, consistent with the distributed repository developed earlier in the book, deeper information can be imagined as **relationally distributed across the underlying substrate**. Individual atoms and structures may carry local information, while persistent cosmic state is encoded across relationships, fields, geometry, or other deeper degrees of freedom rather than depending upon one particular piece of matter.

In computer terms, **matter is not the hard drive containing the only copy of the cosmic database**.

Matter is one participant in a much larger informational architecture.

How Could the Brain Decode It?

This is the most speculative part of the proposal.

A brain contains roughly 86 billion neurons connected through an extraordinarily complex electrochemical network. Thoughts correspond to large-scale patterns of activity across this network. A tiny physical influence does not need to contain an entire thought by itself; theoretically, it might only need to alter which existing neural pathway becomes active at a critical moment.

Imagine a complex system poised between several possible solutions. Years of study have already created almost all the necessary connections. One relationship remains missing.

Within the Cosmic Architecture, a subtle external informational influence could act less like **uploading a textbook** and more like **providing the final key that unlocks an existing structure**.

The sequence becomes:

Years of Preparation
→ Highly Structured Neural State
→ Subtle Informational Coupling
→ Previously Separate Patterns Connect
→ Sudden Conscious Insight

The individual might experience the result as an intuition, inspiration, dream, mathematical insight, or sudden realization. But receiving an insight would not make the insight automatically correct.

A mathematician must still prove the equation. A scientist must test the hypothesis. An inventor must build the device. A philosophical or spiritual insight must withstand reason, experience, and ethical scrutiny.

The **downlink supplies a possibility; the physical world supplies validation.**

Gravity May Be One Layer, Not Necessarily the Whole Answer

The Cosmic Architecture therefore does not need to claim that ordinary gravitational waves alone explain inspiration.

A more sophisticated possibility is that **gravity provides one observable clue that spacetime itself is dynamic and capable of propagating structured disturbances**, while the actual cosmic communication layer could involve deeper field relationships or higher-dimensional structures discussed elsewhere in this book.

In that sense, gravity becomes less like the entire cosmic Internet and more like evidence that **the fabric on which such an architecture would operate is not inert**.

The model therefore proposes a hierarchy:

Dynamic Spacetime → Possible Deeper Information Layer → Prepared Biological Interface → Insight

Established physics supports the first element.

The remaining elements constitute the speculative extension explored by the *Cosmic Architecture*.

And that is exactly where a speculative model should operate: **not by contradicting known physics, but by asking what additional architecture might exist beyond what present observations have yet established.**

5. Gifted Minds as Receptive Catalyst Nodes

Throughout history, extraordinary individuals have appeared whose discoveries fundamentally redirected human civilization. In popular mythology, these figures are often depicted as isolated, superhuman geniuses who built the world out of nothing by themselves.

The *Cosmic Architecture* provides a far more accurate, systems-based perspective: Gifted minds are prepared, highly receptive antenna nodes.

[HIGHER GOVERNANCE TIER] ──► System Update / Knowledge Downlink
│
▼ (Spacetime Gravity Wave / Field Resonance)
[GIFTED MIND (Catalyst Node)] ──► Receives Perception & Translates Downlink
│
▼ (Presents Common Sense & Unified Goal)
[SCIENTIFIC COMMUNITY] ──► Beautiful Minds Collaborate, Validate & Evolve Discovery

The True Role of the Gifted Mind

A gifted mind does not create the universe's mathematical rules out of thin air, nor do they replace the hard work of the broader scientific community.

Instead, a gifted mind acts as a **Catalyst Node**:

1. **Unlocking Higher Perception**: Their neural architecture and deep preparation allow them to tune into a cosmic downlink, perceiving underlying structural relationships in nature, mathematics, and ethics that were previously invisible behind the Knowledge Firewall.
2. **Bridging Perception to Common Sense**: They translate that high-dimensional intuition into localized human language, mathematical equations, or physical concepts.

HISTORICAL KNOWLEDGE DOWNLINKS	
Srinivasa Ramanujan	Raw mathematical theorems received directly from intuitive downlinks, later validated by the global math community.
Isaac Newton	Synthesized gravity and calculus, uniting natural philosophy into a single mathematical framework for all scientists.
Nikola Tesla	Envisioned fully functional AC motors and wireless field principles in complete visual mental flashes.
Albert Einstein	Perceived curved spacetime through thought experiments, giving physics a unified goal for the next century.

3. **Unifying the Community of Beautiful Minds**: They present this new perspective to the broader scientific and engineering community, turning an abstract breakthrough into a clear, universal goal.

We must never belittle the thousands of brilliant scientists, engineers, researchers, and builders who surround these breakthroughs. Gifted minds provide the initial spark of perception, but it is the **collective intelligence of the scientific community**—the thousands of dedicated, beautiful minds testing, refining, building, and engineering—that evolves that initial spark into global civilization.

5. Historical Examples: Receptive Antennae in Action

When we examine history's most profound scientific and mathematical leaps, the signatures of Over-the-Air Knowledge Infusions and community integration become interesting within the proposed Cosmic Architecture model.

A. Srinivasa Ramanujan: Direct Mathematical Downlinks

The legendary Indian mathematician Srinivasa Ramanujan had virtually no formal training in pure mathematics, yet he produced thousands of novel identities, infinite series, and modular equations that were decades ahead of his time. When asked how he arrived at these astounding formulas, Ramanujan maintained that they were revealed to him directly in visions as fully formed mathematical truths. Within the Cosmic Architecture model, Ramanujan's extraordinary intuitive process can be interpreted metaphorically as something resembling a receptive antenna, receiving raw mathematical code from the cosmic repository and presenting it to the world's mathematicians to verify, prove, and expand into modern number theory.

B. Sir Isaac Newton: Unifying Physics into a Universal Goal

When Isaac Newton developed calculus and formulated the universal law of gravitation, he did not work in total isolation. By receiving the intuitive perception that the force pulling an apple to the ground is the exact same force governing the orbits of planets, Newton provided a magnificent "common sense" framework. He handed the scientific community a unified mathematical goal, allowing generations of astronomers, mathematicians, and engineers to collaborate and build the industrial world.

C. Nikola Tesla: Visualizing Field Dynamics

Nikola Tesla explicitly described receiving complete, fully operational technological designs in sudden mental flashes. Before building his alternating current (AC) motors and high-frequency transformers, Tesla could mentally view, operate, and test the machinery in vivid detail. His gifted mind tuned into the underlying electromagnetic field substrate, downlinking hardware concepts that transformed human electrical engineering.

D. Albert Einstein: Re-imagining the Fabric of Space

Albert Einstein's thought experiments (*Gedankenexperiments*) allowed him to visualize what it would be like to ride alongside a beam of light. This intuitive perception led to Special and General Relativity—revealing that space and time are unified into a dynamic, curved spacetime fabric.

Einstein did not build the atomic age alone; his profound perception catalysed thousands of physicists, astronomers, and engineers to test his predictions, leading to modern GPS, astrophysics, and quantum technology.

6. The Two Paths of Cosmic Guidance

To maintain clarity across our architecture, we distinguish between two different types of cross-layer guidance deployed into our sandbox:

Gifted minds and scientific pioneers arrive periodically to expand human capacity, hand us better conceptual tools, and elevate civilization's baseline. Avatars arrive during rare existential crises to restore moral balance (Dharma) when technical power threatens to outpace ethical wisdom.

Neither Gifted Pioneers nor Avatars will save humanity from itself. Pioneers present the expanding perception; the scientific community builds the technology; Avatars remind us of the moral compass; but humanity's ultimate destiny remains strictly in its own collective hands.

PART VI - BUILDING THE COSMIC ARCHITECTURE

Looking Back:

In **Part V**, we discovered that physical matter is not empty, featureless stuff, but a live, highly encrypted partition of a cosmic database. We saw how atoms carry deep stellar history, how instincts query ancient biological repositories, and how higher knowledge patches stream across spacetime to prepared neural receivers.

Moving Forward:

Part VI takes these information and geometry primitives and uses them to construct the operational laws of moral and social reality. We redefine personal identity beyond biological hardware, examine why humanity carries unbound moral agency as Earth's supreme steward, and resolve the classic objection against cosmic intermediate hierarchies. We analyse the Earth Project model, the Cosmic Knowledge Firewall, the deterministic mechanics of Karma, and the rare cross-layer interventions of Avatars. Finally, we resolve the "Clock-versus-Cause" problem of astrology, revealing how celestial transits function as synchronized system audits.

Chapter 21

Personal Identity Beyond Biology

Building on the node framework from Chapter 4, this chapter tackles a major flaw in how many people view life: the belief that a human being is just a temporary biological machine that completely disappears when the body dies.

If identity were purely about biology, two genetically identical twins or two factory-identical laptops would remain the same throughout their existence. But common sense and basic systems engineering show us that hardware is only half the story. A ten-year-old laptop is not defined by its plastic shell; it is defined by the unique collection of files, software, photos, and history stored inside it.

A strictly physical account of personal identity raises a difficult philosophical question by trying to define you only by your physical "hardware" (your body) while completely ignoring your "software" (your accumulated history). You are not just a collection of atoms. To describe who you truly are, you must look at your complete **State Vector**: your memories, moral choices, character, emotional growth, and the relationships you've built over a lifetime.

[BIOLOGICAL MATERIALISM] ──▶ Identity = Physical Body (A temporary container that decays)

│
▼ THE PARADIGM SHIFT
│

[COSMIC ARCHITECTURE] ──▶ Identity = Soul State Vector (Your permanent, saved history)

How This Connects to the Cosmic Architecture

In the **Cosmic Architecture**, nothing meaningful in the universe is treated as disposable garbage. If the cosmos spent 13.8 billion years creating fine-tuned physics, chemistry, and consciousness, wiping out all of a person's life experiences the moment their heart stops would be terrible system design.

Instead, the *Cosmic Architecture* operates on a simple principle: **The physical body is just a temporary terminal, but your Soul State Vector is permanently saved.**

- **Life as a Real-Time Update:** Your life is not just a random survival test. Every decision you make—every act of kindness, courage, greed, or honesty—acts like a software update that changes your internal coordinates in the cosmic network.
- **The Shared Truth Across Traditions:** This is why many world religions and spiritual traditions point toward a deeper reality beyond the temporary physical body. Jainism describes the purification and liberation of the soul; Buddhism emphasizes the cultivation of compassion and freedom from attachment; Hindu traditions explore karma, rebirth, and spiritual development; Christianity emphasizes transformation of character, the soul, and accountability before God; and Islam teaches that earthly life is temporary, that the soul continues beyond physical death, and that human choices carry moral consequences before God. Their theological explanations differ significantly, but they share a broad principle relevant to the Cosmic Architecture: **human existence cannot be reduced simply to the physical body; our choices, character, and spiritual condition carry a deeper significance.**
- **Karma as System Logic:** Karma is not a magical judge in the sky handing out rewards and punishments. It is simply the natural law of conservation in a state-space system. Your past choices create your current starting point, and your present choices shape where you go next.

When physical death occurs, the local 3D body breaks down, but your complete **State Vector** remains saved in the deep Information Architecture of the universe—ready to be preserved, evaluated, or carried forward into the next phase of existence.

Family Sub-Graphs: Love, Debt, and the "Watching Parent" Myth

When a parent passes away, a comforting human sentiment often arises: *"They are up in heaven, watching over me."*

While this captures the enduring warmth of grief and memory, the *Cosmic Architecture* reveals a far more structural truth. When a parent's local biological body expires (node termination), their physical presence closes, and their accumulated state record is finalized for the next cycle. They are not sitting on a cloud, endlessly watching the daily micro-details of 3D life.

Yet, the profound connection between parents, siblings, and children does not vanish into thin air. It persists through the **Family Sub-Graph**.

```
[ PARENT NODE TERMINATION] ──► State Record Finalized & Archived
                 |
                 ▼ (Relational Legacy / Memory)
[ CHILD NODE RUNTIME] ──► High Positive Vector Weight
 (Love, Care, & Respect Active)
```

1. The Positive Loop: Unconditional Family Love

When genuine love, care, respect, and deep emotional bonding exist within a family, that shared warmth carries an exceptionally high **positive vector weight**. Importantly, this works in both directions. A parent's genuine care adds positive state to the parent while contributing protective weight to the child; a child's genuine love and care similarly strengthens the child's state while contributing protection toward the parent.

Within the *Cosmic Architecture*, close family relationships—particularly between parents and children—carry special relational parameters because of their unusually deep developmental and emotional significance. Genuine acts of care and selfless love between parent and child can therefore be conceptualized as carrying **five times the ordinary positive**

protection weight. This 5× value is an internal parameter of the proposed model rather than a scientifically measured quantity.

Even after a parent's physical lifecycle ends, years of care remain embedded in the child's memories, values, emotional resilience, and character. Within the deeper metaphysical layer of the model, the accumulated **protection weight** of that relationship also remains part of the continuing relational state. This offers another interpretation of why people often feel that deceased parents are still "watching over" them: the parent need not literally hover nearby as a ghost; the positive state generated through years of love continues beyond the temporary biological relationship.

Protection does not mean that nothing difficult can ever happen. Rather, within the model, it acts as one positive weighting among many variables influencing the node's wider life trajectory.

2. The Neutral / Negative Loop: Family Friction and Carried State

The same special relational parameters can carry **neutral or negative weight**. Within the reincarnation layer of the *Cosmic Architecture*, unresolved negative relational state may carry forward from a node's previous iterations and reappear within close family relationships. If the unresolved state lies primarily on the parent's side, the parent may encounter difficult behaviour or negativity through the child; if it lies primarily on the child's side, the child may encounter corresponding difficulty through the parent. The relationship becomes a shared recalibration environment in which both nodes are given another opportunity to change their state.

This does not mean that suffering proves guilt or that harmful behaviour should simply be tolerated as "karma." Each node remains responsible for its present choices, and positive actions can begin changing the relationship's trajectory.

3. Siblings and the Wider Family Protection Network

These parameters are not limited to parents and children. **Sibling relationships can also carry positive, neutral, or negative protection weight.** Two siblings may possess unresolved negative state from previous iterations while simultaneously generating new positive protection through genuine love, sacrifice, loyalty, and care in the present life.

The family therefore operates as a small relational network rather than a single parent-child connection. Different edges can carry different weights at the same time:

Love & Care → Positive Protection
Neutral Relationship → Low/Neutral Protection
Conflict & Unresolved State → Negative Protection
Changed Behaviour → Recalibration

Within the *Cosmic Architecture*, family relationships therefore carry unusually strong parameters because parents, children, and siblings repeatedly influence one another's development. **Past state may establish the starting coordinates, but present choices continue to modify them.**

The Structural Takeaway

Family is not a random biological coincidence, nor is it a magical bond of floating spirits. A family is a localized **Relational Sub-Graph** where soul nodes gather to process past debts, express earned love, and refine their collective character.

Whether a family relationship brings profound joy or deep friction, every act of patience, forgiveness, and care actively purges old technical debt—upgrading the node's state vector for the chapters yet to come.

Chapter 22

Why Humanity is the Supreme Species on Earth and Bound by Deeds

1. The Disparity of Agency: Fixed Design versus Boundless Self-Transformation

Compared with humans, most animal behaviour is more strongly constrained by evolved biological drives and immediate environmental pressures. Many animals nevertheless display learning, social intelligence, empathy, problem-solving and behavioural flexibility. Within the Cosmic Architecture, the distinction is therefore one of degree and moral scope rather than an absolute absence of choice.

An eagle, a lion, a shark, or a bee executes precisely what it was biologically designed to do. A lion does not sit under a tree and contemplate whether eating a gazelle is morally questionable; it acts on metabolic necessity and predatory instinct. An ant colony builds intricate subterranean cities, but no ant ever decides to reform the colony's social structure, invent a new language, write poetry, or question the nature of the cosmos. Within the model their behaviours are bounded, predictable, and fixed by evolutionary design. They cannot be held morally accountable for their actions because they possess no capacity to choose against their design.

According to Cosmic Architecture model, human beings represent a fundamentally supreme species on Earth because we alone have been endowed with **unbound moral agency and self-transformative intelligence**.

Humans are dynamic, self-evaluating, emotional, and self-improvising nodes. We possess the unique ability to evaluate our own thoughts, distinguish good from bad, choose against our immediate physical desires, and deliberately alter our behaviour. But this supreme capacity comes with a profound double-edged sword:

- **The Capacity for Heightened Good and Extreme Destruction:** Humans can exhibit unparalleled compassion, building hospitals, sacrificing their lives for strangers, and seeking universal harmony. Simultaneously, humans can be cunning, clever, brutal, manipulative, and destructive—capable of waging wars and altering planetary biospheres.
- **Bound by Deeds (*Karma*):** Because animals have no moral choice, they generate no moral karma; they execute natural design. Humans, however, are **bound by deeds**. Because we possess the freedom to choose between construction and destruction, every choice, intention, and action generates a binding state update on our soul record—dictating our future trajectory across the multi-tiered cosmic architecture.

2. Humanoids Across the Multiverse and Higher Spatial Tiers

This supreme status is not localized strictly to planet Earth. The *Cosmic Architecture* suggests that Earth is not a lonely, accidental anomaly. Across the vast observable universe and higher-dimensional multiverse layers, the "humanoid" or conscious moral template is replicated and assigned specialized tiers:

- **Physical Humanoid Domains:** In our physical universe and parallel physical realms, humanoid species exist at various evolutionary stages. Some may be far more advanced than humanity on Earth, having mastered higher-dimensional physics, emotional integration, and systemic harmony.
- **Graduated Multiverse Tiers:** The multiverse operates through tiered access layers. Earth represents a dense physical training ground—an air-gapped sandbox where conscious entities experience biological mortality and physical limits, learning how their choices directly shape reality.

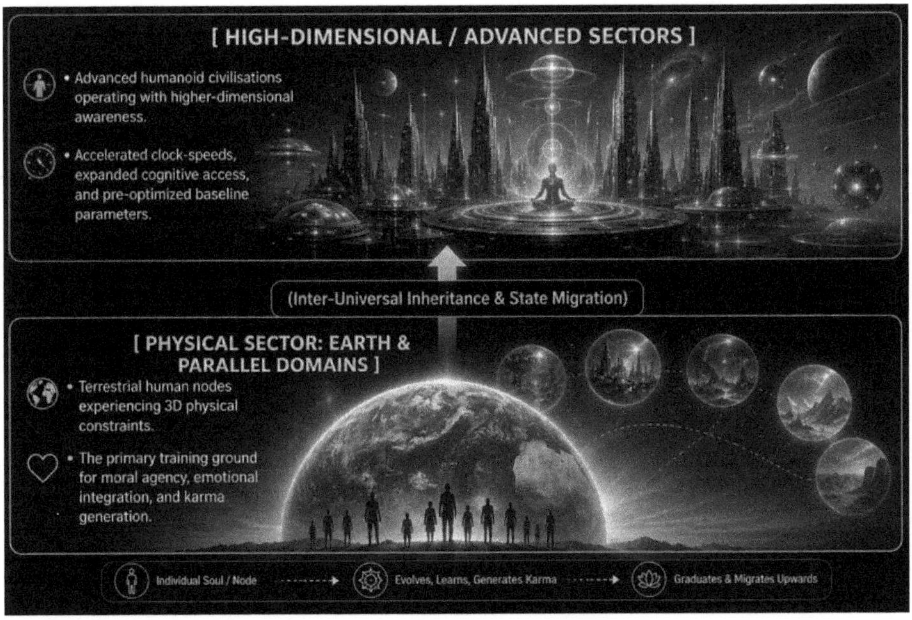

3. Why Does Nature Need Animals if Humans Are the Supreme Species?

If humans represent the supreme conscious species on Earth, a fundamental question arises: *Why does nature need the animal kingdom at all? Why not populate the planet exclusively with humans?*

If a world were populated entirely by billions of human nodes without animals, the planet would feel cold, sterile, emotionally vacant, and ecologically unviable. Animals serve three vital, irreplaceable roles within the Cosmic Architecture:

A. Ecological Orchestration and Biosphere Stability

Animals are the living gears of planetary ecology. They pollinate crops, turn soil, disperse seeds, regulate plant growth, balance microbial populations, and cycle nutrients across land and oceans. Without animals, Earth's biosphere would rapidly collapse, rendering the planet uninhabitable for human life. Animals execute the underlying orchestration rules that maintain the physical stage upon which human life unfolds.

B. Evoking Compassion: The Moral Mirror

Just as parents feel instinctual love, protective care, and compassion toward their helpless children, animals provide a living canvas that evokes **compassion, empathy, and stewardship** in human beings.

Animals test the human soul. How a human node treats a helpless animal—whether with gentle guardianship or senseless cruelty—reveals the true moral state of that individual's soul record. Animals act as a moral mirror, challenging supreme intelligence to express itself as caring stewardship rather than tyrannical exploitation.

C. Nutritional Sustenance and Preventing Self-Destruction

Nature balances humanity through both vegetarian and non-vegetarian dietary pathways. Across human history, animals have formed an essential link in the planetary food chain. The presence of animals preserves ecological interdependence and historical food security.

Chapter 23

Distributed Administration: Operational Scope and Intermediate Governance

Building on the network and hierarchy primitives established in Chapter 4, this chapter examines cosmic administration at the operational and functional level.

In popular discussions of cosmology, religion, and philosophy, governance is often reduced to a rigid, oversimplified dilemma: either a single supreme deity manually controls every falling leaf, collision of atoms, and personal thought, or the universe is a cold, mechanical accident completely devoid of direction. When ancient spiritual traditions introduce intermediate levels of guidance—what various cultures describe as *devas*, angels, lokapalas, or systemic guardians—popular culture frequently caricatures them as mythological courtiers or bureaucratic middle managers.

When viewed through modern systems engineering and organizational architecture, intermediate administration reveals an entirely different logic. It is not an arbitrary pyramid of dominance, nor is it an emergency delegation strategy designed to relieve an exhausted creator. It represents **distributed operational scope, localized stewardship, and the enablement of authentic, autonomous participation across an interlinked cosmic network.**

1. The Operational Tier: Beyond Centralized Micromanagement

In large-scale engineering and organizational design, large-scale resilient systems commonly distribute operational responsibilities rather than routing every micro-level event through a single central controller. If a global electrical grid attempted to route the power switch of every individual living-room lamp directly through one central generator in a distant capital city, the system would immediately fail. It would suffer from massive delays, network congestion, and vulnerability to catastrophic collapse if that single central point encountered an error.

To create stability, resilience, and efficiency, modern systems separate responsibilities across distinct operational layers:

- **The Foundational Baseline:** The core framework that establishes the universal rules, physical boundaries, and fundamental energy laws.
- **Regional Systems & Local Stations:** Autonomous substations, local routing hubs, and regional dispatchers that manage stability, balance loads, and resolve localized fluctuations within their specific geographic sector.
- **Individual Endpoints:** Everyday devices—such as home appliances or smartphones—that operate independently on the ground without requiring real-time instructions from the national power grid for every minor action.

In the *Cosmic Architecture*, the intermediate governance tier (Tier 2) functions precisely as this regional operational layer. The ultimate source of reality does not need to manually steer every ocean wave, force every chemical bond, or script every human decision. Instead, intermediate conscious entities—beings whose awareness, moral maturity, and systemic capability have evolved across immense spans of deep time—are entrusted with operational stewardship over specific planetary clusters, stellar systems, and regional sectors of space.

2. Core Functions of Intermediate System Stewards

Rather than possessing magical, arbitrary powers that defy natural law, entities operating within the Tier 2 Governance layer fulfill specific, highly structured architectural functions within their assigned environments:

Maintaining Regional Environmental Balance

Just as wildlife rangers and environmental scientists monitor a massive national park—ensuring water sources remain clean, animal populations remain balanced, and forest fires are contained before they consume the entire region—intermediate cosmic stewards maintain baseline stability across physical sectors. They monitor environmental resilience, prevent anomalous disruptions from crashing delicate planetary conditions, and oversee the steady operation of planetary sandboxes where developing life evolves.

Summarizing Data and Monitoring Systemic Health

Developing physical worlds produce immense volumes of localized interaction data. Rather than overwhelming higher tiers with raw background noise, intermediate administrators monitor aggregated health reports—tracking species-level moral trajectories, ecological balance ratios, and systemic stress indicators (as detailed in Chapters 5 and 19). They observe whether a civilizational project is growing in wisdom or drifting toward self-destruction.

Enforcing Boundary Isolation

Developing species inside Tier 1 physical embodiment operate under strict isolation protocols (Chapter 25). Intermediate stewards act as boundary guardians, ensuring that primitive sandboxes remain insulated from unearned interference while preventing external advanced species from disrupting the local evolutionary process.

Facilitating Targeted Course-Corrections

When a planetary sandbox drifts toward premature, irreversible collapse due to severe civilizational failure, regional administrators coordinate with root protocols to deploy targeted recalibrations—such as the intuitive knowledge infusions or moral avatar frameworks explored in Chapters 20 and 27.

3. Hierarchy as the Engine of Growth and Participation

Why would an architecture delegate authority rather than automating every process through rigid, unbreakable mechanical physics?

The answer lies in the fundamental purpose of the cosmic system: **the evolution of consciousness through active, responsible participation.**

Consider a classroom where a teacher solves every math problem on the board, writes every essay for the students, and never allows anyone to attempt an experiment on their own. The students would never develop intellectual strength, confidence, or genuine capability. True education requires setting up a safe classroom, defining clear rules, and progressively granting greater freedom as the students demonstrate understanding and discipline.

Within the *Cosmic Architecture*, hierarchy is not an oppressive pyramid; it is the progressive expansion of stewardship:

- **Permissions Follow Moral Alignment:** Operational scope is never granted based on raw ambition or desire for power. It expands strictly in proportion to an entity's demonstrated alignment with non-violence, truth, systemic balance, and selfless care.
- **Stewardship Over Dominance:** In human political hierarchies, rank is frequently exploited for control and personal gain. In cosmic hierarchy, higher rank signifies carrying a heavier burden of care, responsibility, and protection for developing beings lower on the network graph.
- **Preserving the Integrity of Choice:** By distributing governance across localized tiers, the architecture ensures that developing beings experience the genuine consequences of their decisions. The local sandbox remains an authentic testing ground where moral agency carries real causal weight.

4. Administrative Safeguards: Authority Without Tyranny

In human history, whenever power is delegated, it often leads to corruption, arrogance, and systemic abuse. How does a cosmic administration maintain integrity across millions of regional sectors without degenerating into tyrannical empires?

The system enforces automated architectural guardrails:

- **Strict Scope Limits:** Intermediate stewards can only influence parameters within their authorized operational domain. The foundational meta-laws of reality—such as the fundamental balance of causality, karma, and free will—remain locked beyond their reach.
- **Continuous State Logging:** Every administrative action generates an unalterable record within the deeper Information Architecture of reality. No action is hidden, and no misuse of authority goes unaccounted for.
- **Automated Demotion Protocols:** If an intermediate entity develops spiritual pride, abuses its authority, or causes systemic harm to developing worlds, the system executes an automated recalibration—demoting the entity back to lower physical constraints (such as physical biological embodiment) to relearn humility and empathy through material vulnerability (as explored in Chapter 35).

These safeguards ensure that intermediate governance remains a dynamic, self-correcting organism rather than an entrenched bureaucracy. Power is continuously balanced by accountability, and authority remains tied directly to service.

Chapter 24

The Future of Humanity and the Cosmic Project Model

1. Entrusted Autonomy: Why No One Is Coming to Save Us?

A common human tendency during global crises is to look upward and expect a divine intervention or cosmic savior to rescue humanity from its own mistakes. When wars break out, when biospheres degrade, or when political chaos spreads, people ask: "Why doesn't God step in and fix this?"

In the *Cosmic Architecture*, the answer is grounded in entrusted self-governance.

Humanity has been deliberately allowed to evolve, organize into societies, establish political systems, write laws, and distinguish between right and wrong. We operate within a cosmic framework where the Ultimate Source or higher administrators do not micromanage our day-to-day choices. We are not toddlers in a playpen being constantly watched by an overprotective parent; we are an autonomous, self-learning species entrusted with the stewardship of our own planet.

Therefore, blaming God, karma, or the cosmos for human suffering is an evasion of responsibility. If we wage nuclear wars, ruin our environment, or allow greed to destroy our social fabric, we will finish ourselves. The consequences are entirely ours.

2. The Earth Project: Shut Down or Upgrade?

In large-scale engineering and organizational management, projects are continuously evaluated based on performance:

- **The Failed Project:** A project plagued by corruption, internal conflict, bad architecture, and zero progress is eventually decommissioned and shut down.
- **The Successful Project:** A project that master's its challenges, innovates, and achieves systemic harmony receives expanded

resources, evolves to new stages, and its accumulated lessons learned are packaged and transferred to future initiatives.

Humanity on Earth is currently an active Cosmic Project.

If we spend our energy fighting over who is the dominant superpower, squandering resources on warfare, and destroying our environment, the Earth Project will fail and terminate under its own accumulated corruption.

However, if humanity chooses to evolve—overcoming narrow political rivalries to focus on long-term survival—our potential is vast. By focusing on discipline, science, and moral maturity, humanity can:

- Preserve Earth's climate and biosphere for as long as physically possible.
- Develop technologies to control natural disasters, deflect asteroids, and mitigate seismic threats.
- Construct orbital megastructures, space habitats, and off-world colonies.
- Lay the engineering and biological foundations to terraform an "Alternate Earth" over the next thousand years, ensuring the survival of human consciousness even if our primary planet eventually expires.

Within the model, achieving this would mark the Earth Project as a successful developmental phase, with its accumulated state, technological lessons, and conscious wisdom imagined as being preserved and carried forward into wider cosmic inheritance.

Chapter 25

The Cosmic Knowledge Firewall & Non-Interference Protocol

Across modern astronomy and astrobiology, the silence of the night sky remains one of the most profound unanswered puzzles. As explored in Chapter 31, the observable universe contains trillions of galaxies, billions of ancient star systems, and unimaginable physical resources. Mathematically, technological civilisations should have had ample time to expand, communicate, or leave visible engineering megastructures across space.

Yet when humanity trains its radio telescopes, optical observatories, and infrared sensors toward the stars, we encounter an unsettling, unbroken silence.

Standard science often interprets this silence through the lens of absence or physical impossibility: perhaps intelligent life is exceptionally rare, perhaps interstellar distances are an insurmountable wall, or perhaps technological civilisations inevitably destroy themselves during their industrial infancy.

The *Cosmic Architecture* reframes this silence entirely.

Within the Cosmic Architecture model, the silence of the cosmos can be interpreted not as evidence of loneliness, but as a possible structural property of a managed, permission-based environment.

The Moral Sandbox: Why Developing Species Are Isolated

In software systems engineering, when developers build and test new application logic, they do not execute untested code directly inside the production database. Doing so risks corrupting stable data, introducing unrecoverable errors, or causing systemic platform failure.

Instead, developers create an **isolated sandbox environment**.

A sandbox is air-gapped from production. It runs under strict resource constraints, local logging protocols, and parameter boundaries. Inside the sandbox, the application is given full freedom to run its logic, encounter bugs, crash, and attempt recovery—without endangering the broader platform.

Humanity has been endowed with unbound moral agency, abstract reasoning, and high-dimensional state logging. We have been given full freedom to organize into civilisations, establish legal systems, discover physical laws, and choose between constructive stewardship and destructive conflict.

However, because our species is still in its developmental phase—frequently trapped in short-term geopolitical rivalries, resource exploitation, and ecological shortsightedness—we operate under a strict **Cosmic Knowledge Firewall.** Organizations of all sizes use firewalls to block unauthorized incoming traffic and restrict unwanted traffic from leaving their networks.

Role-Based Access Control (RBAC) in the Multiverse

In enterprise platform security, access to sensitive systems is governed by Role-Based Access Control (RBAC), meaning that users receive permissions according to their roles, responsibilities, and authorized level of access. An intern starting on their first day is not granted root-level administrator privileges to the central database infrastructure. Access is matched strictly to demonstrated capability, responsibility, and operational role.

The *Cosmic Architecture* enforces RBAC across high-dimensional state-space through three structural tiers:

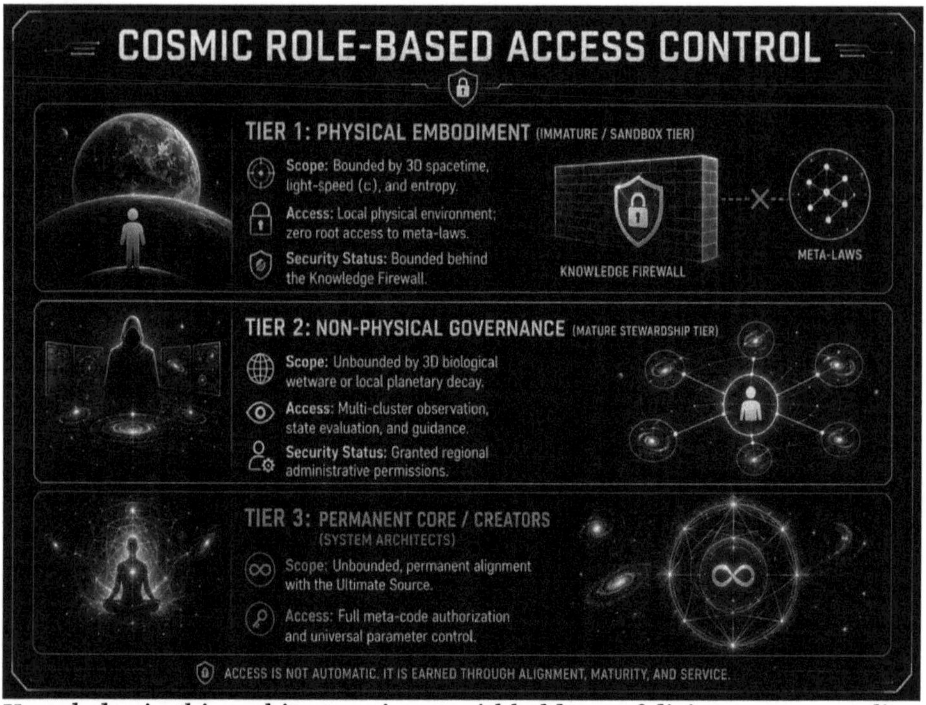

Knowledge in this architecture is not withheld out of divine secrecy or elitism. It is restricted by **systemic compatibility and safety**.

If a developing species were granted root-level access to high-dimensional energy manipulation, zero-point spacetime physics, or universal state editing before mastering its own internal moral alignment, the outcome would be instantaneous self-destruction. Giving advanced interstellar technology to an ethically immature civilization is the equivalent of handing a live nuclear warhead to a toddler in a nursery.

The firewall exists to protect the sandbox from its own unearned power.

The Cosmic Non-Interference Policy: Why Guidance Is Subtle

A common human question arises during times of global crisis: *"If higher intelligences, devas, or cosmic administrators exist, why don't they step in openly, stop our wars, heal our diseases, and hand us clean energy technologies?"*

The answer lies in the **Non-Interference Protocol**.

If an advanced Tier 2 civilization were to land on Earth in broad daylight, erase our diseases, and solve our energy crisis, human independent evolution would immediately stall. As explored in ***Chapter 11***, we would become dependent pets in a cosmic zoo, copying answers we had not earned rather than building genuine systemic wisdom.

Furthermore, as established in ***Chapter 27***, whenever cross-layer intervention does occur within the *Cosmic Architecture*, it is **moral rather than technological**.

Throughout history, extraordinary figures—such as Jesus, the Buddha, the Tirthankaras, and ancient sages—did not arrive with blueprints for space engines or nuclear reactors. They arrived with moral state updates: *compassion, non-violence, truthfulness, self-discipline, and humility.*

They provided the software framework needed to help human nodes govern themselves, without violating the Non-Interference Protocol or shattering the local sandbox rules.

Passing the Systemic Review: The Future of Humanity

The Cosmic Knowledge Firewall is not a permanent prison sentence. It is a **conditional security threshold**.

If humanity fails to master its moral agency—squandering its planetary resources on endless warfare, environmental degradation, and unchecked greed—the Earth Project will eventually hit a self-destruction threshold. As examined through the lens of Astro biological filters in **Chapter 23**, the local sandbox node will fail, its unfulfilled state logs will be archived, and the broader universe will move forward without us.

However, if humanity chooses to evolve—mastering scientific discipline alongside deep moral restraint, establishing global cooperation, and taking responsible stewardship over our biosphere—the parameters shift.

Within the model, increasing moral maturity alongside technical power could correspond to a gradual lowering of the Cosmic Knowledge Firewall, allowing deeper understanding or wider contact to become possible. This is a philosophical mechanism of the framework rather than an established scientific process.

The framework interprets the apparent cosmic silence as a possible mirror of developmental maturity. On this view, technological detection alone may not be sufficient for meaningful contact; a civilization might also need to demonstrate that it can manage powerful technologies without destroying itself. This remains a speculative interpretation of the Fermi question.

What Kind of Civilisation Must We Become?

The biggest barrier to joining the cosmic community is not space travel—it's maturity. Before humanity can explore the stars, we must prove we can safely govern our most dangerous technologies (nukes, AI, bioengineering). Long-term survival requires moral wisdom to match our technological power; otherwise, species destroy themselves before ever reaching the galaxy.

Question to Consider

If an advanced interstellar civilisation were monitoring Earth today, would they view humanity as a mature species ready for open partnership, or as a volatile sandbox node that must remain strictly behind the firewall for the safety of the broader ecosystem?

Chapter 26

Karma as State, History and Consequence

Among all the ideas discussed across the world's ancient religions and modern philosophical traditions, few have generated more persistent debate or widespread misunderstanding than the concept of *karma*. In popular culture, karma is frequently reduced to a simplistic, supernatural reward-and-punishment system—a cosmic scoreboard where good deeds are eventually rewarded with good luck and bad deeds are punished with hardship. Sceptics, on the other hand, often dismiss the idea entirely as primitive superstition, viewing the world as a purely random collection of physical coincidences where actions carry no lasting moral weight.

The reality described by the *Cosmic Architecture*, however, is far more subtle, rational, and interesting than either of these simple descriptions.

The word *karma* literally translates to **action**. At its fundamental level, the principle of karma states that every action produces consequences, even if those consequences are not always immediate, obvious, or linear. A simple seed planted in the soil today may not mature into a fruit-bearing tree for decades. A single word of genuine encouragement spoken to a struggling child can quietly shape that child's self-worth, confidence, and career choices for the rest of their life. A breakthrough discovery made by an isolated scientist in a laboratory can improve human health and medicine for generations to come. Conversely, an unthinking act of cruelty or violence can plant seeds of fear, resentment, and trauma that continue to ripple through families and communities long after the original event has passed into history.

Human civilization already understands and relies upon this exact principle across every major field of scientific inquiry. Actions create consequences, and no event in the physical universe occurs in complete isolation.

- **Psychology** examines how early life experiences, habits, and emotional trauma continuously shape future human behaviour and perception.
- **Sociology** investigates how cultural attitudes, historical decisions, and institutional norms influence generations of citizens.
- **Economics** tracks how individual daily consumer choices and policy shifts ripple outward to reshape global markets.
- **Ecology** demonstrates that introducing or removing a single species within an environment can alter an entire planetary ecosystem.

Different scientific disciplines naturally use different technical vocabularies, but they all acknowledge one universal baseline: the interesting question is not *whether* actions create consequences, but *how far* those consequences extend across time, space, and consciousness.

The *Cosmic Architecture* begins with this exact same observation: **every action changes the baseline state of reality**.

To visualize how this state change operates across a larger system, imagine dropping a small stone into the centre of a perfectly calm mountain lake. The initial splash lasts for only a fraction of a second, yet the resulting circular ripples travel outward across the water far beyond the original point of impact. Most of those tiny waves eventually become far too small for the unaided human eye to notice, but they continue interacting with the water's surface, shifting floating leaves, altering currents, and merging with countless other waves across the lake.

Human intentions and choices behave in a remarkably similar way within the fabric of existence. A simple smile or act of patience alters another person's emotional state in a moment of stress. Seen through the lens of systems engineering, karma is not an invisible judge sitting in the clouds keeping score of human points. It is the continuous, automated, and natural law through which our past choices define our current starting state, and our present actions shape the landscape of possibilities available for our future. Nothing is lost, nothing is arbitrary, and every choice contributes to the ongoing evolution of the whole.

The Threshold Override: Mid-Life Recalibrations and Multi-Cycle Carryover

In the systems framework of the *Cosmic Architecture*, karmic logging operates through weighted parameter calculations rather than simple arithmetic. Actions that inflict intentional physical, financial, or emotional destruction on other nodes carry exponentially higher negative vector weights.

When a node accumulates extreme negative karma within a single physical lifespan, the system enforces two automated execution protocols:

1. Mid-Life Real-Time Recalibration (The Circuit Breaker)

If a node's accumulated negative karmic score crosses a critical safety threshold *before* its physical lifecycle ends, the system executes an immediate mid-life recalculation. The node's active environment shifts within the same lifetime—manifesting as abrupt ruin, severe isolation, mental collapse, or legal containment (e.g., prison). The severe percentage of negative karma overrides past positive reserves, forcing the node to face immediate structural friction.

2. Multi-Lifecycle Debt Persistence

When a node's negative karmic footprint is unimaginably high, the debt cannot be processed or rebalanced within a single 80-year human iteration. The master state vector retains this weighted debt tag across successive cosmic deployments. As a result, the node requires significantly more physical lifecycles than average to clear its state noise, rebuild its moral baseline, and qualify for higher-tier multiverse access.

Why Does the System Allow Extreme Negative Karma?

A fundamental question arises: *If the cosmos continuously logs state vectors and maintains systemic equilibrium, why does it allow a human node to execute extreme negative karma in the first place? Why not step in and block destructive actors before harm occurs?*

The answer lies in two core architectural principles:

THE TWO PILLARS OF FREE-WILL EXECUTION

NON-OBSTRUCTION PROTOCOL	CIVILIZATIONAL SELF-CORRECTION
• Prevents forced automation. • Preserves real moral agency • Choices carry genuine weight.	• Humanity learns by experiencing the consequences of its choices. • Stress-tests build better laws.

1. The Non-Obstruction Protocol

If the cosmic architecture manually blocked every bad choice, genuine moral agency would cease to exist. The local sandbox would lose its purpose as an evolutionary training ground and degenerate into a forced simulation. Free choice exists at the input stage; karma and vector routing operate deterministically at the output stage.

2. Civilizational Self-Correction & Cosmic Experimentation

Humanity is an active, self-learning development project. By allowing conscious nodes to experience the raw outputs of their choices, civilization learns to build law, order, morality, and justice systems through its own historical trials. Dark actors act as "red-team stress tests" that force humanity to write better laws, create human rights frameworks, and cultivate moral discernment. The cosmos observes and evaluates these dynamic trajectories, using the accumulated telemetry to refine future universe deployments.

Chapter 27

Avatars as Cross-Layer Interventions

Throughout human history, virtually every major spiritual tradition has described extraordinary figures who appeared during periods of profound crisis or systemic shift.

- **Hinduism** speaks of *Avatars*—such as Rama and Krishna—divine manifestations descending to restore *Dharma* (cosmic balance and law).
- **Christianity** centres on Jesus, viewed as the incarnation of divine grace entering physical history.
- **Buddhism** describes the Buddha as an awakened teacher who discovered and taught a path beyond suffering.
- **Jainism** honours the *Tirthankaras*, enlightened beings who established ford-makers to guide souls across the ocean of worldly rebirth.
- **Islam**, rather than teaching divine incarnation, describes guidance as revelation from the transcendent God (Allah), delivered through prophets and messengers, including Muhammad—an example of cross-layer communication without God becoming human.
- **Other Traditions** describe prophets, sages, saints, and enlightened messengers who fundamentally redirected the trajectory of human civilization.

While these traditions differ in theology and culture, they share a striking observation: at pivotal moments in history, individuals appear whose influence extends far beyond a single lifetime, altering the moral and cultural baseline of humanity for millennia.

[Cultural History View] ──► Studies Social Influence, Leadership & Historical Context

|
▼ THE ARCHITECTURAL QUESTION
|

[Cosmic System View] ──► Asks How Higher Tiers Intervene Without Breaking System Rules

Historians study the societies in which these figures lived. Psychologists examine their leadership and charisma. Anthropologists explore how cultures preserve their teachings. None of these disciplines can determine whether a person was cosmically appointed or divinely inspired—they can only study what is physically observable.

The **Cosmic Architecture** asks a deeper, systemic question:

If reality normally develops through natural laws, free choice, and gradual evolution, why do exceptional beings appear at specific critical junctions in history?

1. The Balance Between Autonomy and Guidance

Earlier chapters introduced the **Cosmic Non-Interference Policy**. If intelligent life is to mature, it cannot simply be handed every answer. A civilization that receives unlimited knowledge without developing the moral wisdom to use it will inevitably destroy itself.

Human education follows this exact principle. Teachers guide students without doing their homework for them. Parents gradually increase a child's independence rather than solving every personal problem throughout their lives. If the cosmos constantly micromanaged every human decision, it would defeat the core purpose of free will and evolutionary development.

Yet, complete non-interference creates a dangerous vulnerability: *What happens if a civilization drifts toward catastrophic self-destruction before completing its evolutionary cycle?*

When a civilization reaches this critical tipping point, the Higher Governance Tier within the model, this can be represented as—deploying an

Avatar or moral framework as a root-level system patch to recalibrate the species without revoking its free agency.

This is not a violation of system logic; human systems balance autonomy and intervention in the exact same way every day:

- **A driving instructor** lets the student control the vehicle but grabs the wheel if a fatal crash becomes imminent.
- **A doctor** allows the body to heal naturally whenever possible but performs emergency surgery when a condition becomes life-threatening.
- **Software engineers** allow automated pipelines to run independently, but step in with a targeted code patch when a critical bug threatens total database corruption.

Good systems balance operational independence with rare, high-priority interventions. Within the Cosmic Architecture, reality can be modelled as operating under a comparable architectural balance.

2. The Avatar as a Root-Level System Patch

In enterprise software engineering, when a running application develops severe algorithmic drift or corrupted logic that local microservices cannot fix from within, administrators do not destroy the entire platform. They execute a **hotfix**—injecting clean, root-level code directly into the active system to recalibrate baseline parameters.

In the language of *cosmic architecture*, an Avatar (derived from the Sanskrit *Avatara*, meaning "descent") is an **injected root-level patch**:

[Higher Governance Tier] ──► Injects Pure System Code (Avatar) ──► [Local 3D Earth Sandbox]
(Devas / Supreme Source) (Recalibrates Drift / Patch)

1. **Local System Drift:** Over centuries, human free choice can lead to severe moral decay, institutional corruption, or unchecked violence that threatens to end the project prematurely.
2. **Systemic Inability to Self-Correct:** Because local nodes (humans) are locked inside the 3D sandbox and influenced by local corruption, they often lack the perspective or power to reset the system on their own.
3. **Cross-Layer Injection:** The Higher Governance Tier executes a targeted cross-layer intervention. A soul-vector carrying root-level authorization, pure moral clarity, and higher-dimensional wisdom is deployed into a local human body.
4. **Recalibrating the Baseline:** The Avatar does not use raw force to destroy free will. Instead, they re-establish foundational laws, inject new moral algorithms, and demonstrate higher ways of living—rebalancing the local sector without crashing the system.

3. Why Interventions Are Exceptionally Rare

If cross-layer interventions are possible, why don't Avatars appear every year to solve all human problems?

In system design, frequent manual overrides destroy system integrity. If a system administrator must step in every five minutes to fix user errors, the underlying automation is broken.

[Frequent Manual Overrides] ──▶ Destroys User Free Will & System Integrity

│
▼

[Rare Strategic Injections] ──▶ Preserves Free Will While Preventing System Crash

- **Preserving System Integrity:** The cosmos allows human nodes maximum freedom to experience the consequences of their choices, generate karma, and learn through trial and error.
- **High Threshold for Intervention:** A cross-layer intervention occurs only when local conditions threaten total system failure—

when moral or physical drift reaches a critical threshold that would lead to premature project termination.

Avatars do not arrive to do humanity's work for us. They arrive to **reset the compass**, leaving behind timeless frameworks, moral codes, and spiritual roadmaps. Once the patch is applied and the system is recalibrated, the Avatar steps back, allowing humanity to continue its evolutionary journey using its own free choice.

4. Summary: The System's Safety Valve

Are historical spiritual figures merely mythological characters or ordinary charismatic leaders?

When viewed through our speculative architectural framework, they are something far more profound: **they are the safety valves of the cosmic system.**

They demonstrate that while our 3D world operates under automated physical laws and human free will, it is not an abandoned experiment. Within Cosmic Architecture, they can be interpreted as an actively monitored, masterfully architected network—one where autonomy is respected, growth is earned, and emergency intervention steps in whenever necessary to ensure that the light of consciousness is never permanently extinguished.

Chapter 28

Shani, Astrology and the Clock-versus-Cause Problem

CAUSE, CLOCK OR COINCIDENCE?
Across millennia, human cultures have interpreted planetary cycles as corresponding with patterns in individual lives and human history. Whether such correlations exist beyond coincidence, selective interpretation, and psychological bias remains scientifically unestablished. In classical Indian astrology (*Jyotisha*), few cosmic principles evoke as much reverence and trepidation as **Shani** (associated with the planet Saturn) and its major planetary epoch, the *Shani Dasha* (or its intense seven-and-a-half-year transit, *Sade Sati*).

Popular tradition often interprets Shani as a punitive cosmic judge—a distant physical planet actively beaming malevolent rays or bad luck toward specific individuals to inflict suffering.

Modern scientific scepticism rightly rejects this literal mechanism: gravity drops off with the inverse square of distance, and the electromagnetic forces exerted by Saturn on a newborn infant are far weaker than those emitted by ordinary household electronics.

The **Cosmic Architecture** does not require distorting Newtonian physics or electrodynamics to entertain celestial correlations. Instead, it reframes the entire phenomenon by resolving a fundamental conceptual confusion: **The Clock-versus-Cause Problem.**

1. The Clock-versus-Cause Problem

To explore how planetary movements might correlate with human experience without relying on magical force fields, consider how synchronized maintenance schedules operate in engineering.

The Synchronized Service Analogy

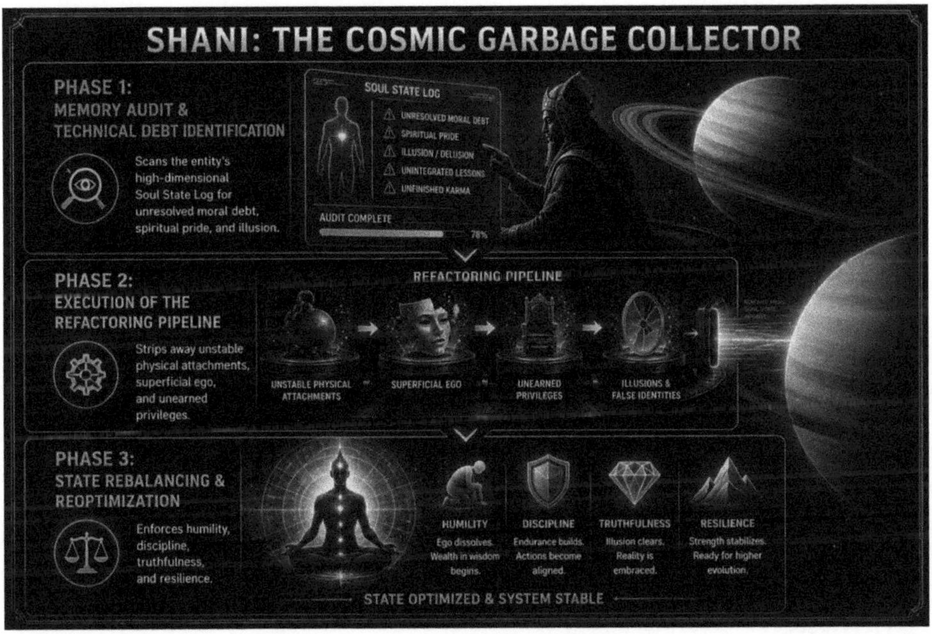

Imagine a city water treatment plant whose central controller triggers a full filter overhaul every 30 years. Attached to the plant's main intake pipe is a large mechanical counter dial that completes one full revolution across that 30-year span. Concluding that the dial's needle is hostile or physically forcing the plant to restrict water flow is a category error. The needle does not cause the maintenance; it simply marks a pre-scheduled service threshold designed to flush out accumulated sediment and ensure long-term stability.

Shani's 30-Year Orbital Timestamp

In Jyotisha, Shani (Saturn) takes approximately 29.5 years to complete a single orbit around the Sun, spending roughly 2.5 years passing through each zodiacal sign. Its movement therefore provides a slow and highly predictable celestial cycle.

Importantly, Saturn's movement is **global, while its traditional astrological interpretation is individual**. Everyone alive on Earth exists under the same astronomical position of Saturn at a given time, but periods such as Sade Sati are calculated relative to an individual's natal Moon. Therefore, the same Saturn transit is not interpreted as producing the same developmental phase for every person.

Within the Cosmic Architecture, this distinction can be represented as a **global clock with node-specific triggers**:

[GLOBAL CELESTIAL CLOCK]
Saturn follows its predictable ~29.5-year orbital cycle
↓
[CURRENT SYSTEM TIMESTAMP]
Same astronomical Saturn position for all Earth-based nodes
↓
[INDIVIDUAL STATE MATCHING]
Each human node has a different natal baseline, life history, karmic state and current state vector

↓
[NODE-SPECIFIC TRIGGER]

Only compatible state conditions correspond with a particular developmental or review cycle

In this interpretation, Saturn does not need to cause events, nor does every person experience the same process simultaneously. The celestial position functions more like a **shared system clock**, while each individual node responds according to its own configuration and state.

Think of thousands of scheduled processes running on the same computer system. They all read the same system clock, but they do not all execute at the same moment. One process may be scheduled for 10:00, another for 10:30, and another for midnight. **The clock is global; the trigger condition is local.**

Position as a System Indicator, not a Causal Force

Saturn is not emitting targeted "negative energy rays" across space to ruin a person's life. Rather, its physical position in the sky functions as the **visible minute hand on the master cosmic clock**.

When Saturn transits specific coordinates relative to an individual's natal baseline, within the Cosmic Architecture model, such a Saturn transit can be interpreted symbolically as indicating a scheduled Garbage Collection and Memory Audit for that specific human node. Within the model, such periods are interpreted as opportunities for reassessment, discipline and restructuring. Difficulty would not necessarily imply personal failure or moral deficiency; ordinary causality, biological events, social conditions and the free choices of other people would continue to operate.

2. Shani as the Universal Garbage Collector & Refactoring Engine

In software engineering, large enterprise systems continuously accumulate temporary files, unoptimized memory allocations, technical debt, and lingering state corruption. To prevent system crashes, administrators run automated **Garbage Collection (GC)** routines and scheduled **refactoring pipelines**.

Garbage collection is rarely pleasant for running applications—it pauses background threads, frees up bloated memory, terminates unoptimized processes, and forces the code to run strictly within allocated resource limits.

In the *Cosmic Architecture*, **Shani represents the systemic Garbage Collector and Runtime Auditor.**

Stripping Technical Debt: During a *Shani Dasha* or *Sade Sati* period, the cosmic runtime engine executes a forced refactoring pipeline on the conscious node. False pride, unearned power, superficial attachments, and toxic behavioural patterns are forcibly purged.

- **The Illusion of Punishment:** The process feels severe or painful only because human nodes naturally resist losing comfortable illusions. If a software thread fights against memory clearance, it experiences latency and errors. If a human node embraces discipline, humility, and structural truth during this cycle, the refactoring succeeds, resulting in long-term resilience, wisdom, and promotion to higher scopes of responsibility.

3. Cascading Network Triggers & State Vectors

How does a celestial "clock interval" manifest as real-world events, such as a sudden job loss, structural re-alignment, or financial hardship?

It operates through **cascading network triggers** across the high-dimensional state-space:

1. **Systemic Synchronization:** The conscious entity's state vector reaches a point where its accumulated historical actions (karma) require evaluation.
2. **Causal Network Cascades:** Instead of requiring supernatural intervention, the model imagines changes propagating through

graph relationships. Small upstream shifts in neighbouring nodes, economic environments, social relationships, or health conditions can naturally cascade downstream, creating circumstances that require the individual to reassess existing structures and choices. The circumstances may change, but the individual's response remains open, preserving moral agency and free will.

4. Keeping the Framework Grounded: Testing the Clock Model

To ensure *The Cosmic Architecture* remains grounded in reason rather than unfalsifiable pseudoscience, we must be honest about where different models stand:

1. **Direct Physical-Force Model (Unsupported):** There is no established scientific evidence that Saturn's ordinary gravitational or electromagnetic influence produces individualized astrological effects. The Cosmic Architecture therefore does not rely on such a mechanism.
2. **The Psychological Model (Symbolic Utility):** Viewing planetary cycles as archetypes for self-reflection can be helpful psychologically, but it treats astrology as a human metaphor rather than a cosmic reality.
3. **The State-Space Clock Model (Open Research Horizon):** Our model proposes that celestial positions act as synchronized indicators of systemic state evaluations.

[DIRECT PHYSICAL-FORCE MODEL] → Saturn directly causes individualized events → Unsupported
[PSYCHOLOGICAL MODEL] → Archetypal / reflective framework → Symbolic Utility
[STATE-SPACE CLOCK MODEL] → Synchronized system indicator → Open Research Horizon

How to Prove the Clock Model Scientifically

To distinguish a genuine cosmic clock from mere coincidence or human imagination, any real test must clear two simple hurdles:

- **Filter Out Human Bias:** It must eliminate the *Barnum Effect* (vague predictions that sound true for anyone) and *Confirmation Bias* (remembering the few times a prediction came true while ignoring all the times it failed).
- **Statistical Clustering:** It must demonstrate that major, structural life events—such as unexpected career shifts, health recalibrations, or forced periods of isolation—happen during specific planetary transits at rates significantly higher than random chance.

If rigorous testing shows no reproducible correlation beyond chance, there would be no empirical reason to treat the planetary positions as indicators of individualized life events. If strong, reproducible correlations survived independent testing and rigorous controls, however, they would justify further investigation into why those correlations occur. The Cosmic Architecture's synchronized-clock model would then become one possible hypothesis to test against competing explanations—not a conclusion established by correlation alone.

PART VII - MULTIVERSE, COSMIC INHERITANCE AND UNIVERSE LEARNING

Looking Back:

In **Part VI**, we built the moral and administrative architecture of our local sandbox. We mapped karma as dynamic state recalibration, explored how human choice updates the soul log, and established that cosmic governance operates through distributed scope, non-interference firewalls, and rare system hotfixes.

Moving Forward:

In **Part VII**, we zoom back out to the macro scale to answer the ultimate questions of cosmology. Why is the universe so absurdly, unimaginably vast? Why build a magnificent creation only to allow it to end in entropy? We examine the cosmos as a multi-tenant cloud platform, where universes exist as air-gapped tenants separated by permission firewalls rather than physical miles. We explore how universes inherit pre-set blueprints, learn from past cycles, and execute surgical moral ascension. Part VII resolves the Vastness Paradox—revealing that space is segmented, dynamic expansion is air-gapping, and cosmic termination is harvesting rather than loss.

Chapter 29

The Multi-Tenant Cosmos: Isolation, Firewalls, and Higher Governance

We see stars and galaxies scattered across billions of light-years, and we naturally assume that our universe is a standalone, isolated physical room.

If someone mentions the word *multiverse*, popular culture paints a picture of floating soap bubbles separated by vast physical distances, or science-fiction portals allowing advanced civilizations to build ships and travel between parallel worlds. The Cosmic Architecture explores a very different possibility.

Cosmic Architecture suggests that we are not alone, but we also cannot simply build a fast rocket or discover a clever equation to travel to another universe—no matter how many millions of years a civilization must evolve. The multiverse is not separated by physical miles; it is separated by **structural permission boundaries and cosmic multi-tenancy**.

1. The Multi-Tenant Architecture: The Cloud Model of the Cosmos

The Non-Technical Metaphor Mapping Key

Technical Azure Concept	**Plain-English / Real-World Metaphor**	**How it maps to the Multiverse Model**
Invisible Azure Infrastructure (Hypervisor / Data Centre)	**The Physical Foundation / Central System Engine**	The underlying cosmic framework that hosts all reality. It manages power, space, and laws without any user seeing the hardware.

Technical Azure Concept	Plain-English / Real-World Metaphor	How it maps to the Multiverse Model
Azure Tenant (Isolated Environment)	**An Air-Gapped Private Office / Isolated Apartment**	A self-contained universe. It shares the background engine with other universes, but has its own front door, walls, and privacy.
Source Code / Underlying Scripts	**The Blueprint / Master Operating System**	The fundamental mathematical laws written by the creators. Users experience the output (gravity, light) but cannot read or alter the source code.
Provisioned Resources (VMs, Databases)	**Pre-Furnished Equipment (Stars, Planets, Atoms)**	The physical assets deployed inside a universe for conscious beings to interact with.
Role-Based Access / Network Firewalls*(Chapter 25)*	**Security Clearance / Digital Pass cards**	The invisible boundaries that keep civilizations in their assigned "sandbox" until they qualify for higher access.

To understand how parallel universes can exist right alongside our own without crashing into each other, modern cloud computing provides a far better model than floating soap bubbles: **Multi-Tenant Architecture**.

Think of a massive, modern apartment building or a secure digital banking platform:

- **The Invisible Background Engine (The Cloud Infrastructure):** In the background sits a giant, invisible foundation that supplies power, security, heating, and structural support to the entire building. The residents living inside the apartments don't see the underlying pipes, electrical wiring, or structural blueprints—they simply enjoy the living space.
- **The Isolated Tenant (The Universe):** Each apartment represents a **Tenant Universe**. Two families live right next door to each other, sharing the exact same physical building foundation. Yet, because of solid walls and locked front doors, Family A cannot see into, walk into, or interfere with Family B's apartment.
- **The Source Code (The Underlying Blueprint):** The master creators set the background rules—configuring the apartment layout, lighting, and utilities. The residents use the furniture, kitchen, and appliances (our planets, stars, and atoms), but they cannot alter the master architectural blueprint hidden inside the walls.

Why This Model Explains Parallel Realities

When applied to cosmology, our universe is simply one **Tenant Workspace** hosted on a vast Cosmic Infrastructure.

The creators provisioned our tenant with a rich set of local resources—stars, galaxies, planets, and chemical elements—and set the background physics in motion. As human beings, we interact with these pre-furnished physical resources every day. However, we cannot peek behind the curtain to edit the "source code"—the underlying mathematical laws and high-dimensional boundaries that isolate our tenant from neighbour worlds.

Within the Cosmic Architecture, parallel universes are not imagined as being billions of light-years away in ordinary empty space. Instead, the model pictures them as sharing a deeper cosmic foundation while remaining separated by structural boundaries.

2. The Strict Isolation Between Universes

Why would a supreme architect or higher governance hierarchy build such strict isolation between universes?

A. Preventing Systemic Civilizational Clashes

If the physical boundaries between universes were permeable, aggressive, highly developed civilizations from ancient, multi-billion-year-old universes could easily invade and harvest younger, developing universes. To prevent predatory dominance and unearned destruction, the cosmos enforces strict **tenant isolation**. Each universe acts as a protected sandbox where species can evolve, make moral choices, and experience their own developmental cycle without external interference from older, vastly superior realms.

Feature	System Function
Local Independence	Species evolve naturally without interference from outer universes.
Hard Firewall	Prevents physical travel or signal breach across tenant boundaries.
Root Governance	Only Higher Administrative Tiers oversee and manage all tenants.

B. The Limits of Local Technology

A common sci-fi assumption is that if a species becomes smart enough, it can "breakthrough" into other dimensions. But in a managed architecture, local tools cannot override root infrastructure. Just as a program running inside a web browser cannot rewrite the host operating system's kernel, a biological or technological entity inside a 3D universe cannot bypass the **Cosmic Knowledge Firewall** as established under the Knowledge Firewall in **Chapter 25**. Access to multi-tenant structures is strictly restricted to the Higher Governance Tier (what ancient traditions call Devas, System Managers, or Higher Administrators).

The higher administrators do not micromanage every detail inside each tenant. They monitor aggregated telemetry, evaluate net moral output versus wasted noise, and manage cross-tenant migrations when local lifecycles conclude.

Chapter 30

Cosmic Inheritance and Universe Learning

The Big Bang occurred, matter gathered to form stars and planets, life evolved, and eventually the universe will slowly run out of energy and end in a dark, cold void known as **Heat Death**. In this familiar view, every universe exists in absolute isolation—it has its own beginning, its own lonely history, and its own conclusion. Once it ends, the story simply stops.

The *Cosmic Architecture* invites us to explore a much richer possibility: **What if a universe is not a disposable container running out of steam, but an active environment in an unimaginably long process of learning, refinement, and continuity?**

1. The Principle of Cosmic Inheritance

Human civilization thrives because every generation inherits the discoveries, inventions, and lessons of the past. We do not force young students to re-invent basic mathematics every century, nor do we require engineers to discover electricity from scratch every generation. Instead, young minds begin where previous generations left off. Books, libraries, and schools preserve discoveries, so civilization continually moves forward rather than repeatedly starting over from zero.

If human society relies on inheritance to avoid wasting time and energy, would a master cosmic system operating across billions of years ignore this basic principle?

Over vast timelines, an advanced cosmic system observing countless universes would recognize which physical rules foster long-term stability and which lead to early collapse. It would identify the exact physical settings and environmental conditions that allow life to thrive—using those accumulated lessons directly to set up future creations.

2. How Pre-Set Blueprints Shape Cosmic Deployment

To understand how the cosmos creates a new universe without hand-crafting every single detail, consider how a master architect builds a modern skyscraper or how a developer sets up a complex digital system:

Instead of manually building and configuring every component one by one, a systems designer can create an **automated blueprint**—a foundational set of rules, templates, and parameters. Once that blueprint is defined, automation can use it to repeatedly create and configure complex environments with minimal manual intervention. In modern software engineering, a similar approach is called **Infrastructure as Code (IaC)**. Rather than manually configuring every computer, server, network, or database individually, engineers define reusable templates and supply different values where needed for each environment. The same underlying blueprint can therefore produce many different configurations without rebuilding everything from scratch.

In the *Cosmic Architecture*, this master blueprint concept is the core engine behind **Cosmic Deployment**:

[COSMIC MASTER BLUEPRINT] —▶ Sets Core Rules & Physical Constants
|
▼
[RUNTIME UNIVERSE] —▶ Chemistry, Stars, & Life Unfold Automatically

1. **Setting the Constants (The Parameters):** When a new universe is launched, the Cosmic Architecture does not manually position every star or hand-craft every living organism. Instead, it establishes the **foundational rules**—such as the exact strength of gravity, the speed of light, and the laws of atomic chemistry.
2. **Automated Unfolding (Emergence):** Because these starting parameters are pre-optimized based on lessons learned from previous universes, the environment unfolds naturally. Stars ignite, planets form in stable orbits, and chemistry naturally builds self-replicating biological life.

3. **No Micromanagement Required:** Just as a well-designed blueprint allows a city to grow organically without the architect directing daily traffic, pre-set cosmic rules allow infinite variety and complexity to blossom while keeping the universe structurally stable.

3. Granular Governance & Multiverse Transitions

When considering how the cosmos manages reality, a critical question arises:

What happens if humanity on Earth is destroying its biosphere through war and greed, while intelligent species on thousands of distant worlds across the same universe are living peacefully and morally? Does the cosmos destroy the good planets along with the bad ones?

Per the Cosmic Architecture model, the answer is a clear **no**. The *Cosmic Architecture* does not use blunt, indiscriminate tools. Just as a cloud administrator can isolate an infected file while moving healthy databases to a secure, premium environment, the cosmos governs by moral trajectory through **Surgical Sorting and Multiverse Transitions**:

Pathway A: Individual Node Re-Embodiment (Micro Level)

At the individual level, when a conscious entity (a soul node) completes its physical lifespan, its local 3D terminal shuts down:

1. **Node Shutdown:** Physical hardware dissolves.
2. **State Logging:** The node's accumulated character, choices, and karma are saved to the Cosmic Archive.
3. **Cross-Tenant Provisioning:** The central engine evaluates the saved record and re-embodies the soul node into a matching environment in another universe sector.

Pathway B: Civilizational Ascension & Local Decay (Macro Level)

At the macro level, the architecture can be imagined as sorting entire planetary clusters, stellar arms, or regional sectors based on collective moral alignment.

Targeted Cluster Migration (Moving the Good)

If a planetary civilization passes the **Cosmic Maturity Test**—achieving long-term non-violence, ethical stewardship, and systemic equilibrium—the architecture does not require the species to physically build engines to fly outside their universe.

Within the speculative framework of *The Cosmic Architecture*, the transition can be modelled through **Topological Phase Shifts**:

- **Substrate Re-mapping over Physical Relocation:** The system does not drag billions of tons of matter through 3D space like a towing truck. Instead, the underlying spacetime coordinates of that sector are topologically reassigned to an advanced multiverse tenant.
- **The Seamless Local Frame:** To an observer standing on the surface of an ascending planet, the local physics remain continuous. Gravity functions normally, planetary rotation continues, and the immediate night sky looks consistent. However, the sector has transitioned into a higher-tier baseline where physical decay, thermodynamic exhaustion, and chaotic resets no longer destabilize the environment.
- **Selective Sector Harvesting (Pruning and Retaining):** A galaxy or stellar sector is rarely uniform. Corrupted or unstable regions within a sector can remain bound to the decaying local sandbox to run their natural course of entropy, while refined, coherent sub-clusters are harvested and migrated into matching multiverse domains.

The Optical Horizon: Why Advanced Civilizations Cannot See the Transition

A natural question arises: *If a planetary cluster or galaxy sector is migrated to another multiverse realm, why wouldn't neighbouring astronomers in the old universe spot it instantly with powerful telescopes?*

Within the model, this is explained by the fundamental physics of light and causal boundaries:

- **Causal Decoupling:** In physical cosmology, light travels across spacetime at the invariant speed *c*. When a sector undergoes topological migration into another tenant domain, its photons no longer propagate into the old 3D metric.
- **The Light Barrier & Blackout Effect:** As the migrated sector shifts beyond the local coordinate manifold, incoming and outgoing electromagnetic signals are air-gapped. To an outside observer in the original universe—even one with multi-million-year-old telescope technology—the migrated region appears as an empty void, a dark cloud, or a gravitational relic. Because light cannot traverse tenant firewalls, the old universe can neither receive fresh data nor visually track the migrated system.

Local Sector Decay (Isolating the Bad)

Self-destructive civilizations remain locked in the decaying physical sandbox. The cosmos does not need to send catastrophic punishment out of anger. It simply enforces local cause and effect: unchecked resource extraction, warfare, and systemic disharmony cause the local environment to degrade until the unstable species terminates itself under its own broken logic.

Moral Realism: Morality as a System Credential

Within the Cosmic Architecture, this suggests a cosmos in which goodness, compassion, and harmony are not merely social preferences but possible indicators of developmental readiness for higher levels of responsibility.

The framework treats that connection as a metaphysical proposition, not as scientific proof.

4. Encrypted Memory and Universal Learning: How the Past Shapes Future Creation

Whether an individual planet fails, a local sandbox is decommissioned early, or an entire cosmos lives out its natural lifespan, the *Cosmic Architecture* obeys one fundamental rule: **Nothing meaningful is ever thrown away.**

To a non-technical reader, terms like "archiving data" or "pre-set parameters" might sound like abstract computer jargon. But the way the multiverse preserves history and "learns" is identical to how human civilization improves over time.

Step 1: Compressing History (The Master Summary)

Before any planet or universe is dismantled, the system extracts its net achievements, moral breakthroughs, scientific discoveries, and structural failures.

- **The Everyday Analogy:** Think of an entire 300-page historical encyclopedia being summarized into a single, high-density digital cheat-sheet. You don't need to save every individual scrap of scratch paper used by every student; you preserve the final proven formulas and key historical lessons.
- **The Multiverse Application:** The cosmos compresses the entire story of a sector—every act of love, every moral choice, and every physical law that worked—into a **High-Dimensional State Record** (a master digital blueprint holding the essential lessons of that environment).

Step 2: Safekeeping in the Cosmic Archive

This compressed dataset is transferred out of the local, decaying physical space and preserved within the deeper distributed informational substrate described in Chapter 2.

- **The Everyday Analogy:** When you upgrade to a new phone, you don't throw your old photos, contacts, and apps into the trash. You upload them to a secure cloud account, so your history remains intact even after the old physical phone is recycled.
- **The Multiverse Application:** The **Cosmic Archive** is a logical abstraction over the distributed substrate. It holds the accumulated history of past creations so that cosmic progress is never lost to entropy or decay.

Step 3: Guiding the Next Generation (How a Universe "Learns")

When we say a universe "learns" or "teaches," we don't mean that a universe possesses a human brain or personality. In system design terms, learning simply means **using past experience to build a better future system**.

- **The Everyday Analogy:** When software engineers release Version 2.0 of an operating system, they don't start from scratch like beginners. They look at Version 1.0, fix the bugs that caused crashes, optimize the battery settings, and add better features based on what worked.
- **The Multiverse Application:** Within the model when a new universe or planetary sandbox is launched, it does not receive a copy-pasted duplicate of every old rock and star. Instead, it inherits an **upgraded master blueprint**:
    - **Pre-Optimized Physical Laws:** Fine-tuned gravity and atomic forces that prevent early cosmic collapse.
    - **Faster Paths for Life:** Chemical settings that allow self-replicating life to boot up more smoothly.
    - **Expanded Consciousness Windows:** Environments that allow moral wisdom and intelligence to flourish with less internal friction.

This leads to a fundamental question: *If the Supreme Architecture has already nurtured advanced, multi-billion-year-old civilisations in older multiverse sectors, why start a new project like Earth from scratch? Why go through the long process of staging hydrogen, forging stars, cooling crusts, and waiting for biological cells to evolve?*

In modern technology, engineers do not stop building new platforms just because they have stable, legacy software running. Older platforms, while functional, carry historical constraints. When a system architect launches a new project, they do not duplicate the past; they utilize the accumulated R&D of every previous iteration to deploy a platform that is **faster, more efficient, and structurally superior**.

Within the *Cosmic Architecture*, Earth represents a **Fast-Tracked Showcase Sandbox**—the cosmic equivalent of a living "Disneyland" or a high-density, rich-environment prototype:

1. **Accelerated Deployment Cycles:** Thanks to pre-optimized physical constants (inherited from past universe deployments), Earth transitioned from raw stellar dust to complex, self-aware biological minds in a fraction of the time required by primordial universes.
2. **A Canvas for Richer Complexity and Beauty:** Earth was provisioned with an extraordinary convergence of parameters—liquid water oceans, rich chemical libraries, dynamic seasonal cycles, and complex emotional telemetry. Within this thought experiment, it was engineered to test how deeply art, love, empathy, moral agency, and physical elegance could express themselves within a dense physical sandbox.

The Cosmic Architect does not start a new planet because the old ones failed; the architect starts a new planet to express an even higher tier of harmony, beauty, and conscious complexity made possible only by the lessons of the past.

Chapter 31

Can an Advanced Civilization Breach the Creators?

A fascinating question naturally arises when we model the universe as an evolving, learning architecture:

If an ancient civilization inside a multiverse sector evolves for tens of millions or billions of years, could it eventually acquire enough technological power to breach the system boundaries, hack the root infrastructure, or overthrow the Cosmic Architects themselves?

At first glance, the question seems plausible. Humanity has already learned that intelligence can rapidly convert into dangerous capability. Nuclear physics gave us power over atomic energy before our moral and political systems were mature enough to govern it wisely. Artificial intelligence now presents a similar challenge: technical capacity is advancing faster than institutional wisdom.

However, as established in **Chapter 25**, the cosmos enforces strict knowledge firewalls, access limits, and non-interference protocols. **Chapter 31** further explains why young civilizations are isolated inside vast causal and spatial buffers. This chapter does not need to repeat those firewall principles in full. Instead, it focuses on the deeper question behind them:

Even if a civilization passes the moral tests of maturity, why can it still never bridge the multi-cycle advancement gap between a local universe and the Creator Enterprise that initialized it?

1. The Early Cosmic Sandbox: Trial, Error, and Systemic Resets

Within the framework of *The Cosmic Architecture*, the Supreme Creators did not necessarily deploy a flawless, fully governed multiverse on the first cosmic cycle.

In the earliest primordial universes—the first experimental sandboxes—the architecture may have been less refined. Within the framework, a civilization admitted to higher tiers would have received expanded technological parameters before moral safeguards, causal boundaries, and knowledge filters were fully matured. As intelligence accelerated, those species could have abused high-dimensional energy, destabilized local domains, or attempted to interfere with neighbouring sectors.

In systems terms, those early universes functioned as unstable test environments. When a sandbox became too corrupted to preserve, the Creator Enterprise would not need to negotiate with its internal nodes. It would simply decommission the failed environment, preserve whatever useful lessons were generated, and incorporate those lessons into later cosmic deployments.

```
[ EARLY PRIMORDIAL CYCLES ]
    └── Unrestrained capability growth
          └── Boundary stress / instability
                └── System reset and lesson extraction

[ LATER MATURE CYCLES ]
    └── Inherited guardrails
          └── Built-in causal limits and governance protocols
                └── Sustainable long-term evolution
```

This is not a repetition of the knowledge firewall argument from Chapter 25. Rather, it explains how those firewalls could have emerged: not as arbitrary restrictions, but as inherited civilizational wisdom encoded into later universe structures.

2. Pre-Set Governance and the Multi-Cycle Gap

Later universes—including our own—would therefore inherit pre-set blueprints refined by earlier failures. These blueprints include hard physical limits: invariant light speed, light-cone causality, expanding cosmic distances, entropy constraints, and dimensional separation between local domains.

But the most decisive barrier is not merely physical. It is historical.

A civilization inside our universe might eventually become unimaginably advanced by human standards. It might survive for billions of years, master stellar engineering, understand higher-dimensional mathematics, and communicate across vast cosmic regions. Compared with humanity, such a civilization would appear godlike.

Yet compared with the Creators, it would still be a recently initialized local process.

Human technological civilization is only a few centuries old. An advanced alien civilization might be five billion years ahead of us. But the Creator Enterprise represents an intelligence that has evolved across entire cosmic cycles—perhaps millions or trillions of complete universal iterations. The gap is not simply one of tools or weapons. It is a gap of inherited memory, dimensional processing, substrate access, architectural authorship, and cycle-level experience.

[HUMANITY]
Centuries of technical development
↓
[ADVANCED LOCAL CIVILIZATION]
Billions of years of universe-contained development
↓
[CREATOR ENTERPRISE]
Millions or trillions of cosmic cycles of inherited architecture

A local species can mature within the rules of its universe, but it does not stand outside the process that generated those rules. It can learn to operate inside the sandbox with extraordinary sophistication, but it cannot retroactively become the author of the sandbox's underlying architecture.

3. Why Mature Civilizations Do Not Attempt a System Hack

An advanced civilization that survives for billions of years would have done so precisely because it matured beyond impulsive domination. In the language of this book, it would have passed the Cosmic Moral Test. It would have learned cooperation, restraint, long-range systems thinking, ecological balance, and humility before the scale of reality.

Such a civilization would not behave like an adolescent species intoxicated by its first weapons. It would understand that power without wisdom leads to self-destruction. It would also recognize that attempting to attack the Creator Enterprise is not merely immoral; it is strategically meaningless.

A mature species would have vast local work to do:

- exploring high-dimensional geometry within permitted bounds;
- stabilizing its own planetary and stellar networks;
- collaborating with peer civilizations;
- refining consciousness and moral intelligence;
- preserving knowledge across deep time;
- preparing its own universe's lessons for future inheritance.

From such a perspective, "hacking the Creators" would look like a primitive fantasy—an immature projection of human insecurity onto a reality billions of developmental layers beyond us.

The wiser question is not: *can we overthrow the Creators?*

The wiser question is, *how do conscious civilizations become trustworthy enough to inherit greater scope without corrupting the systems they touch?*

4. The Substrate Boundary: Why the Gap Is Not Merely Technological

Even if a civilization becomes peaceful, brilliant, and morally stable, it still faces a deeper boundary than distance, weaponry, or engineering capability. It faces the substrate boundary.

A local civilization develops inside one universe. Its instruments, mathematics, bodies, and cognition are shaped by that universe's physics. Even if it learns to manipulate spacetime, harvest stars, communicate across galaxies, or interact with hidden dimensions, its entire scientific framework remains downstream of the laws it inherited.

The Creator Enterprise, by contrast, operates at the level where those laws are selected, refined, inherited, and deployed. It does not merely discover physics; it authors the conditions under which physics becomes discoverable.

This is why the breach question cannot be solved by imagining a bigger starship, a stronger telescope, a higher-energy particle accelerator, or a smarter artificial intelligence. Those tools are still local instruments running inside the universe's permitted execution layer. They may reveal more of the sandbox, but they cannot climb outside the architecture that generates the sandbox itself.

A species may eventually understand every accessible layer of its own universe. It may map galaxies, decode quantum fields, and infer the existence of deeper architecture. But inference is not rooting access. The species remains an internal process trying to reason about the higher-order substrate that makes its reasoning possible.

This is the decisive boundary: a local intelligence cannot use the outputs of a system to seize the source-layer that generates those outputs.

For that reason, the multi-cycle gap is not only a matter of age. It is a difference in ontological position:

- the local civilization is an evolved process inside a universe;
- the Creator Enterprise is the higher-order architecture through which universes are initialized and sustained;
- the local species can study effects, but it cannot seize the source-layer that makes those effects possible.

This is why even a civilization billions of years beyond humanity does not become a rival to the Creators. It may become magnificent within its domain, but it remains downstream of the architecture that gave that domain existence, causality, memory, and lawful stability.

5. The Locked Knowledge Horizon

Within the framework, a mature civilization may come to interpret reality through ideas such as inheritance, governance, causal boundaries, and moral evaluation. It may deduce that its universe is part of a larger multiverse architecture. It may even infer that deeper Creator-level intelligence exists beyond the local physical domain.

But inference is not access.

Just as human beings can reason about dimensions they cannot directly perceive, a local civilization can reason toward the Creator Enterprise without reaching its root layer. The final origin remains locked beyond the knowledge horizon.

[HUMAN SANDBOX]
Local moral and technological development
↓
[ADVANCED LOCAL SPECIES]
Billions of years of universe-contained science
↓
[CREATOR ENTERPRISE]
Cycle-spanning architectural authorship
↓

[ULTIMATE ORIGIN]
Locked horizon beyond local cognition

The infinite-regress question — *Who created the Creators? What existed before the first cosmic cycle?* — belongs to the later discussion of divine governance and ultimate origin. As **Chapter 34** explores, this is where 3D biological cognition reaches its hard authorization limit. We may reason toward the structure, but we cannot force access to the first cause.

5. The "Rogue Enterprise" Objection: Why an Internal Species Cannot "Hack" the Source

A practical reader might wonder:

In the real world, a brilliant senior engineer at a tech giant can leak source code, or a nimble startup can grow to disrupt an industry giant like IBM. If a civilization evolves over billions of years—mastering dark matter, dark energy, and space-time bending—couldn't a 'rogue senior species' eventually go neck-to-neck with the Creator Enterprise?

While this sounds intuitive, it mistakes **using a platform's tools** for **owning the platform's foundation**.

1. **Application Features vs. Operating System Substrate:** Bending space-time or harvesting dark energy is not "hacking the code." It is simply using the advanced physical equations built *into* the universe. An engineer using a powerful software program is still working *inside* the program; they do not own the computer hardware running it.
2. **Built-In System Safeguards:** Hyper-scale systems do not rely on simple passwords. They use isolated environments (sandboxes) and strict permission boundaries. The moment an internal process attempts an out-of-bounds action, automatic circuit breakers isolate or reset that specific environment.
3. **The Substrate Barrier:** An advanced civilization operates on borrowed energy, borrowed matter, and pre-compiled physical

> laws provided by the Creators. You cannot use the outputs of a system to overthrow the very framework that generates your existence.

A civilization can become a magnificent, highly advanced steward *within* its universe, but it remains structurally downstream of the architecture that gives it life, intelligence, and physical law.

Transition Note

This chapter has used the substrate boundary only to answer the breach question: *why an advanced local civilization cannot cross from universe-contained intelligence into Creator-level authorship*? The full discussion of the Living Multidimensional Awareness Substrate, higher administrative beings, and the symbolic governance roles represented by Brahma, Vishnu, Shiva, devas, angels, and system stewards is developed later in **Chapter 37** and **Chapter 34**. Here, the point is narrower: the Creators cannot be breached like a distant machine, because every local civilization remains inside the architecture it is trying to surpass.

Chapter 32

The Vastness Paradox: Zone Isolation, Cosmic Horizons, and the Purpose of Lifecycle Resets

When humans look out at the observable universe—a breathtaking expanse spanning over 90 billion light-years containing trillions of galaxies—two obvious, deeply uncomfortable questions arise:

1. *If humanity is one of the few developing species in our sector, why is the universe so absurdly, unimaginably huge? Why would a master cosmic architect waste trillions of light-years of empty space just to host a handful of worlds like Earth?*
2. *And why build such a magnificent, vast creation only to allow it to eventually decay, collapse, or dissolve in the end?*

In traditional astronomy, this overwhelming scale can make humanity feel utterly insignificant—a tiny, accidental speck floating in a cold, indifferent void. In science fiction, it often leads to the naive assumption that space is simply an open ocean waiting for us to build faster starships to conquer it.

To a purely materialist observer, a universe that expands for billions of years only to face eventual Heat Death or destruction looks like absolute futility—like an artist painting a masterpiece only to burn the canvas.

The *Cosmic Architecture* suggests a completely different reality: Space is not wasted, space is sectored along a Multiverse Chronology, and destruction is not loss.

```
[THE DUAL PARADOXES OF COSMIC DESIGN]
|
├── 1. THE VASTNESS PARADOX (Why so huge?)
|   └── Space is Segmented via Multiverse Chronology: Enforces Zone Isolation & Air-Gapped Sandboxes.
|
└── 2. THE DECAY PARADOX (Why let it end?)
```

└── Physical Hardware is Temporary: Harvests State Vectors & Preserves History in the Archive.

1. Space is Segmented along Multiverse Chronology

What our limited 3D physics perceives as vast, empty void separating distant stars is an artifact of our low-dimensional perspective. Our physical universe is a multi-tenant canvas partitioned into distinct operational sectors along a master multiverse timeline:

- **Earth's Nursery Sector:** Earth resides in a localized developmental sector—an air-gapped nursery bounded by the speed of light (c) and physical distance. This isolation gives young species like humanity the freedom to make choices, experience consequences, and generate karma without disrupting mature realms.
- **Overlapping Advanced Sectors:** Meanwhile, distant regions or higher dimensions within our same astronomical universe host advanced civilizations that **overlap directly with Advanced Multiverse Tiers**. While they share our overarching physical framework, their operational coordinates and state alignment grant them access to higher-dimensional networks, bulk gravity, and systemic governance.

Within the multi-tenant fabric of this model, cascading updates do not struggle across light-years of empty void; they propagate through high-dimensional gravitational resonance. Thus, pivotal events—such as comets delivering prebiotic organic code to early Earth—can be imagined not as purely random events, but as localized physical deployments along high-dimensional geometric coordinates

2. Termination is Harvesting, Not Destruction

As established in **Chapter 30**, a physical universe is a temporary runtime sandbox. When a cosmic cycle reaches its natural conclusion, the physical hardware (stars, planets, and dust) dissolves, but the accumulated state records, moral breakthroughs, scientific discoveries, and soul intelligence are compressed, harvested, and committed directly into the Master Cosmic Archive.

The Cosmic Architect does not build a universe to preserve physical rocks forever; the architect builds the environment to harvest the permanent intelligence and moral agency generated within it. The end of a universe is not a tragic defeat, but the successful completion of a major processing run—preparing refined parameters for an upgraded deployment in the next cycle.

The apparent vastness of the cosmos, the invariant speed limit of light (c), and the accelerating expansion of space are not random physical quirks. Within the model, they can be interpreted as automated runtime execution of Zone Isolation, Buffer Networks, and Cosmic Security Protocols.

A. The Myth of the Predatory Multiverse: Refining the Moral Baseline

Before examining *why,* the cosmos enforces zonal separation, we must correct a common sci-fi misconception.
It is tempting to imagine that the multiverse is isolated to protect young worlds from being invaded and destroyed by older, high-tech alien empires. But as established throughout this framework, **a civilization cannot reach the higher tiers of the multiverse through brute force or raw technology alone.**

To survive across millions or trillions of years and gain access to advanced multiverse zones, a civilization *must* pass the **Cosmic Moral Test**. They must master non-violence, environmental stewardship, emotional balance, and systemic harmony. Truly advanced civilizations in the higher realms are not greedy, predatory imperialists looking to conquer younger worlds. They have long transcended greed and physical warfare.
The real security risk flows in the **opposite direction**.

B. The Risk of the Young Node: Why Protocols Are Enforced

Consider how modern human organizations handle system security. An enterprise bank or cloud provider does not implement strict zero-trust security protocols because they expect their senior executive board to launch a ransomware attack. They implement protocols because **untested, unauthorized, or immature users** will inevitably push buttons they don't understand, attempt to bypass access controls, or introduce chaos into stable environments.

Humanity—and other species in their civilizational infancy—represents this exact operational risk:

- **The Hunger for Unearned Power:** Young civilizations develop an intense hunger for advanced energy sources, higher-dimensional physics, and powerful technologies long before they develop the moral maturity to govern them safely.
- **The Vulnerability of Immature Intrusions:** Even if a primitive species cannot truly "breach" an advanced, higher-tier system, an ethically reckless species attempting to harvest zero-point energy or tamper with higher-dimensional fields could destabilize local spacetime, generate catastrophic noise, or disrupt the delicate balance of neighbouring sectors.
- **Preserving System Integrity:** Just as a system administrator locks down root-level directory permissions to prevent an intern from accidentally crashing the central server, the cosmos enforces strict protocol boundaries to protect mature, harmonious zones from the unpredictable chaos of ethically immature worlds.

3. The Mechanics of Separation: How Physics Enforces the Protocols

How does the Cosmic Architecture physically enforce this zonal separation without needing visible walls or giant shields in space?

It bakes the protocols directly into the **laws of physics**.

A. The Invariant Speed of Light (c) as a Bandwidth Cap

In physics, the invariant speed c is the local upper limit for causal signal propagation through spacetime. The Cosmic Architecture then adds a speculative interpretation: that this physical limit also functions as a natural boundary on how quickly a developing civilization can influence distant regions.

B. Accelerated Expansion (Dark Energy) as Dynamic Air-Gapping

One of the greatest mysteries in modern cosmology is why the expansion of the universe is accelerating—a phenomenon attributed to "Dark Energy."

In the architectural model, accelerated expansion can be represented as a form of Dynamic Zone Maintenance. As sufficiently distant regions recede and causal horizons develop, the metaphor imagines an increasing buffer between cosmic sectors. Standard cosmology describes the expansion; the additional idea that it functions to isolate developing civilizations is speculative.

PHYSICAL ENFORCEMENT OF COSMIC PROTOCOLS.	
Speed Limit	Acts as a physical bandwidth cap on causal reach.
Dark Energy	Drives spatial expansion to maintain dynamic airgaps.
Light Cones	Restricts physical interaction to immediate local sectors.

4. Are we Already Inside the Multiverse Web?

When we combine high-dimensional topology with zonal protocols, a profound truth emerges: **We are not sitting in an isolated universe waiting to find the multiverse.** The model asks us to imagine that we may already be living inside a wider multiverse web.

What we call "our observable universe" is simply our assigned **Local Sandbox Zone**—a partitioned, air-gapped environment within the vast multidimensional database cluster.

In this master architecture:

1. **You Cannot "Fly" to the Next Zone:** An advanced rocket or a clever mathematical formula will never allow a species to cross zone boundaries, because the separation is governed by **permission levels, not physical distance**.
2. **Qualification Over Exploration:** A young civilization does not enter higher zones by building bigger engines; it qualifies for

reassignment through **moral evolution, spiritual maturation, and systemic harmony**.

3. **Surgical Re-Embodiment:** When an individual conscious node or a planetary species achieves true moral alignment, the system executes a **Targeted Cluster Migration** upon lifecycle completion—re-allocating the node into a higher-tier, prospering zone within the invisible multiverse network.

5. Advanced Zone Dynamics: Discovery, Diplomacy, and the Cosmic R&D Engine

A crucial question arises when mapping the higher tiers of the multiverse: *Once a civilization passes the Cosmic Moral Test and graduates out of its primitive nursery zone, what happens next? Do all advanced civilizations immediately see one another, live in absolute perfection, or can conflicts still occur?*

A. Contradicting Sci-Fi & The Multi-Century Diplomatic Arc

In the *Cosmic Architecture*, promotion to an advanced zone elevates a civilization's physical and dimensional clearance but finding and integrating with neighbouring worlds remains a deliberate, multi-century journey.

This is where our architectural model fundamentally contradicts popular science fiction:

[POPULAR HOLLYWOOD SCI-FI] —► Instant hyper-drives, dramatic first contact, & impulsive space wars

|

▼ THE ARCHITECTURAL REALITY

|

[COSMIC ARCHITECTURE] —► multi-century stealth, cautious observation, & phased diplomatic integration

- **The Fallacy of Sci-Fi "Instant Contact":** Movies often depict first contact as an impulsive, dramatic event—a giant spaceship suddenly hovering over a skyline or a hyper-drive ship stumbling into an alien fleet, immediately leading to trade or war. In reality, mature species do not rush recklessly into open exposure.

- **Wisdom Born from Historical Caution:** Human history shows the devastating cost of premature contact between asymmetric civilizations—such as the tragic destruction of Native American and Australian Aboriginal cultures following European arrival. Advanced civilizations across the multiverse have long transcended these destructive patterns. Having studied the historical telemetry of thousands of developing worlds across deep time, they carry a far deeper, more sober understanding of the risks of cultural shock and unearned contact.
- **Superior Stealth Beyond Human Imagination:** Because young civilizations like humanity are still in their technical infancy, we tend to assume that finding an alien world is just a matter of pointing a bigger telescope or building a faster radar. We cannot yet fathom how a multi-million-year-old civilization hides itself. They do not merely paint their ships black; they utilize high-dimensional geometry, metric warping, and thermal-signature dissipation to remain entirely invisible within the vast fabric of space until contact is intentionally warranted.
- **The Multi-Century Diplomatic Arc:** When two advanced civilizations finally detect one another, first contact is not a single afternoon event. It unfolds across centuries through a meticulously structured, multi-tiered diplomatic pipeline:

[PHASE 1: PASSIVE OBSERVATION] ──▶ Decades/Centuries of non-intrusive telemetry & frequency monitoring

│
▼

[PHASE 2: INITIAL SYMBOLIC SIGNAL] ──▶ Subtle mathematical exchanges verifying mutual alignment & intent

│
▼

[PHASE 3: GRADUATED DIPLOMACY] ──▶ High-level cultural, scientific, & ethical alignment protocols

│
▼

[PHASE 4: PHYSICAL VISITATION] ──► Controlled exchange & physical integration into the shared zone

1. **Phase 1: Passive Observation (Centuries of Telemetry):** Both nodes quietly monitor each other's emissions and state vectors across decades or centuries to verify mutual ethical stability and peaceful intent.
2. **Phase 2: Initial Symbolic Signal:** Communication begins with subtle, mathematical signal exchanges—verifying that both species share complementary values and respect the rules of the higher zone.
3. **Phase 3: Phased Level 1 Diplomacy:** Delegations engage in long-distance cultural, scientific, and ethical alignment before any physical proximity is attempted.
4. **Phase 4: Physical Visitation & Network Integration:** Only after centuries of proven trust, shared diplomacy and mutual respect do physical visitations occur, slowly weaving two distinct species into a cooperative, high-zone network edge.

By replacing Hollywood's impulsive space-war tropes with a multi-century diplomatic pipeline, the Cosmic Architecture proposes a different measure of advancement: not merely technological speed, but patience, caution, restraint, and diplomatic wisdom across deep time.

B. High-Zone Conflicts, "Might vs. Right," and Governance Guardrails

Does promotion to an advanced zone mean a species becomes infallible, unfeeling angels?

Not necessarily. "Advanced" is a relative term. Beings in higher zones are far superior in wisdom, diplomacy, and technical power compared to 3D human nodes, but they remain evolving entities within their own scope.

- **High-Level Diplomacy Over Raw Destruction:** Primitive species resolve disagreement through physical warfare (blasters and planetary bombardments) because their vision is short-sighted and resource scarce. Advanced species resolve friction through high-dimensional diplomacy, moral alignment, and systemic equilibrium.

- **The Role of Might and Wisdom:** If a disagreement occurs between two advanced nodes of unequal development, the less experienced node naturally defers to the broader scope, deeper telemetry, and established wisdom of the more mature civilization.
- **Systemic Recalibration for Bad Actors:** In stories like *Marvel's* depiction of cosmic entities (or the *Guardians of the Galaxy*), rogue actors occasionally attempt to misuse higher power. In *the Cosmic Architecture*, if a rogue faction or individual within an advanced civilization attempts to misbehave or exploit system rules, the **Higher Governance Tier (System Administrators)** steps in. The system executes surgical recalibrations—revoking permissions, re-assigning coordinates, or demoting the rogue node back to lower physical constraints to relearn foundational lessons.

C. The Prime Directive: Why Young Worlds Remain Invisible

Science fiction fans will recognize this principle immediately: in *Star Trek*, the **Prime Directive** strictly forbids advanced starships from interfering with primitive, pre-warp civilizations.

In documentaries like *Cosmos*, astrobiologists discuss the **Zoo Hypothesis**—the idea that advanced aliens might be actively watching Earth from a distance while keeping themselves hidden so as not to disrupt our natural cultural evolution.

Within the Cosmic Architecture model, this is imagined not merely as a social agreement but as a deep constraint built into the operating rules of the universe:

- **Zero Direct Interference:** Advanced civilizations cannot arbitrarily invade, conquer, or colonize young developmental sandboxes.
- **The One-Way Firewall:** The **Cosmic Knowledge Firewall**(as established under the Knowledge Firewall in Chapter 25) acts as an

impassable airgap. Developing worlds (like Earth) cannot perceive the advanced zones, and advanced zones are blocked from disrupting the local free-will execution of developing nodes.

D. The Ultimate Purpose: The Distributed Cosmic R&D Engine

This resolves the ultimate question: *Why does the Supreme Source require this vast process of advancement, discovery, and higher-tier development?*

If the universe were meant to be a static, pre-scripted machine, a single static realm would have sufficed. The cosmos designed a multi-tenant, evolutionary architecture because it values **novel learning, distributed innovation, and active creation**.

Popular discussions of theoretical physics sometimes ask whether the universe can be understood as an information-processing or learning system. The Cosmic Architecture explores a stronger speculative possibility: advanced civilizations could function as a distributed research-and-development layer within a much larger cosmic process:

- **Exploring Unfathomable Frontiers:** Free from the primitive struggles of disease, physical scarcity, and war, advanced species dedicate their collective intelligence to research—mapping higher-dimensional geometry, pioneering new forms of art and harmony, manipulating zero-point energy, and solving meta-cosmic equations that human brains cannot even formulate.
- **Commits to the Central Cosmic Archive:** The breakthroughs, structural lessons, and moral solutions generated by these advanced civilizational clusters do not vanish. They are continuously logged, summarized, and committed back to the **Master Cosmic Archive**.
- **Upgrading Future Creations:** When new universe deployments are initialized, the Supreme Architecture reads from these accumulated research commits. The lessons learned by today's advanced species directly enrich the pre-set parameters and biological blueprints for tomorrow's creations.

Far from being passive observers, advanced civilisations are active, conscious co-creators—working in partnership with the Cosmic Architecture to expand the boundaries of possibility across deep time.

Summary: The Universal Zero-Trust Architecture
Why is the universe so vast?
It is vast because it is a masterfully architected, multi-zone environment operating under a **Cosmic Zero-Trust Protocol**.
The trillions of light-years of expanding space are not empty waste. They are the dynamic buffer zones designed to allow young species like humanity the freedom to grow, learn, and face the consequences of their choices inside a safe sandbox—without endangering the eternal, prosperous realms of the wider multiverse.

PART VIII - GOD, EXPLANATION AND LIMITS

Looking Back:

In **Part VII**, we unlocked the multi-tenant architecture of the multiverse. We saw how dark energy acts as dynamic air-gapping, how physical constants are inherited as pre-set blueprints, and how mature planetary sectors undergo surgical cluster migration into permanent multiverse domains.

Moving Forward:

Part VIII confronts the ultimate philosophical boundaries of human inquiry. We carefully reconcile science and purpose—establishing why discovering a physical mechanism never disproves underlying design, while avoiding the logical trap of the "God of the Gaps." We address why an infinite source utilizes distributed governance, explore structural polarity (division of labour) across macro-creation, and examine the Demotion Recalibration Protocol for corrupted administrative nodes. Finally, we stand at the edge of the Trillion-Year Enterprise, acknowledging the hardcoded knowledge horizons that prevent 3D biological hardware from processing the root-level origin of the Creators.

Chapter 33

Mechanism vs. Purpose: Why Science and Architecture Coexist

One of the greatest achievements of human civilization is modern science. Astronomy explains how stars are born inside nebulae, geology explains how tectonic plates build mountains, biology explains how species adapt through natural selection, and neuroscience maps the electrical signals firing across the human brain.

Every scientific discovery expands our understanding of nature. The *Cosmic Architecture* fully embraces and celebrates these achievements. It does not compete with science, nor does it ask you to reject proven physical facts. Whenever science provides a well-supported explanation for *how* something works, we accept it completely.

Yet, even after a physical process is fully explained, a fundamentally different kind of question always remains.

Science explains how physical mechanisms operate, but it does not by itself determine whether the laws underlying those mechanisms possess an ultimate purpose. Confusing these two questions has created centuries of unnecessary conflict between science, philosophy, and spirituality.

```
[PHYSICAL SCIENCE]──► Maps the Mechanics (HOW the engine runs)
                  │
                  ▼ DIFFERENT DOMAINS
                  │
[COSMIC ARCHITECTURE] ──► Explores Possible Design/Purpose (WHY
the rules allow it to run)
```

1. Runtime Execution vs. System Design

In modern systems design, there is a clear boundary between the **Runtime Environment** (how code executes step-by-step) and the **System Blueprint** (why the parameters and rules were written that way).

RUNTIME MECHANICS vs. SYSTEM DESIGN	
1. Runtime Mechanics	Natural selection, chemical bonding, gravity, genetics.
2. System Design	The physical constants and math that permit life to emerge.

Consider how this distinction applies across major scientific disciplines:

- **Cosmology & Physics:** Physics explains how gravity pulls floating gas together to ignite stars. But *why* does our universe possess fine-tuned gravitational constants that allow long-lived stars to form instead of collapsing into immediate black holes?
- **Biology & Evolution:** Natural selection explains how species adapt to environmental pressures over generations. But *why* is physical chemistry configured with self-organizing properties that allow dead matter to assemble into self-replicating DNA code in the first place?
- **Neuroscience & Mind:** Neuroscience maps electrical pulses moving across brain tissue. But *why* should physical electrical currents produce first-person, felt subjective experiences like love, beauty, or moral choice?

Science maps the step-by-step execution inside the sandbox. The *Cosmic Architecture* explores the underlying rules, constants, and purposes that made the sandbox possible.

2. Avoiding the "God of the Gaps" Trap

To keep our framework grounded, we must strictly avoid a famous logical trap known as the **God of the Gaps**. Throughout history, whenever humans didn't understand something—like lightning, eclipses, or disease—they prematurely claimed: *"God did it directly!"* As science advanced and discovered natural explanations for lightning and germs, those primitive religious claims collapsed. Using human ignorance as "proof" of God is a terrible strategy.

The *Cosmic Architecture* follows two strict guardrails:

1. **Ignorance is Not Proof:** An unanswered scientific question is never automatic proof of a higher power. Science will continue to solve many of today's mysteries through natural laws, and we welcome those discoveries.
2. **Mechanisms Do Not Eliminate Meaning:** Conversely, discovering a physical mechanism does not prove the absence of design. Explaining how nuclear fusion powers the Sun does not explain why the universe runs on mathematical laws capable of creating suns, planets, and conscious life.

3. Intellectual Humility: Science and Purpose Working Together

The purpose of this book is not to replace science, nor is it to force ancient religious dogmas onto modern physics. The goal is to build a bridge of **intellectual humility**:

[SCIENCE] ──► Gives us Fact, Empirical Rigor, and Technical Mechanics.
│
▼ SYNTHESIS
│
[PHILOSOPHY & ARCHITECTURE] ──► Gives us Meaning, Moral Purpose, and Context.

When we look at the universe through this dual lens, the cosmos ceases to be either a magical, lawless fairytale or a cold, accidental pile of dust. It can be contemplated as a masterpiece of systemic engineering—a dynamic, self-organizing runtime environment governed by precise mathematical laws, where physical mechanisms carry out a far deeper, multi-dimensional purpose.

Chapter 34

Distributed Governance Without Divine Limitation

When people consider the idea of a supreme creator or an ultimate reality, they almost automatically picture a single, central figure managing every single detail directly.

So, when they encounter concepts of lower-tier managers, helper beings, or intermediate administrators—what various world traditions call *devas*, *angels*, or *lokapalas*—an obvious philosophical objection pops up:

If the ultimate source possesses infinite power and infinite awareness, why would any middle management or lower hierarchy be needed at all? Doesn't having assistants prove that the creator gets tired, runs out of processing capacity, or cannot handle the workload alone?

At first glance, this sounds like a strong argument. In human life, delegating work is usually a sign of limitation. A company CEO hires managers because they don't have enough hours in the day, and a hospital adds staff because a single doctor cannot examine every patient at once.

If we project this human limitation onto the cosmos, intermediate administrators look like evidence of weakness—as if God were an overworked executive who needs middle managers to handle the daily paperwork.

However, when we examine how complex, living systems operate, a very different principle emerges: **Hierarchy and distributed stewardship do not reflect limitation; they reflect wisdom, enablement, and the joy of participation.**

1. The Parent, the Teacher, and the Living Garden

Consider how wise parents raise a child.

A parent could easily steer every turn of the bicycle handlebars, tie every shoe, and solve every homework problem for their child. The child would fall less often, but they would also remain permanently helpless. True parental wisdom consists of creating a safe sandbox, setting clear rules, and gradually stepping back so the child can make choices, learn from consequences, and mature into an independent adult.

The exact same pattern appears throughout nature and modern systems:

- **The Forest Ecosystem:** A thriving forest does not require an invisible manager standing over every tree, manually pulling roots through the soil. The foundational laws of biology, sunlight, and rainfall set the environment, while soil networks, plants, and animals interact locally. Order emerges through shared relationships.
- **Distributed Cloud Networks:** Early computers tried to process every task inside one central processor. But as global systems grew, engineers learned that allowing local routers, edge nodes, and regional microservices to manage local traffic made the network vastly more resilient, flexible, and efficient. Engineers didn't distribute the workload because the main computer ran out of energy; they did it because **sharing responsibility is a far better way to build a stable system.**

2. Hierarchy as the Engine of Growth and Stewardship

In the *Cosmic Architecture*, intermediate administrators (*devas* or higher governance tiers) do not exist to solve a staffing problem for an overworked creator. They exist so created beings can **participate, take on responsibility, and grow.**

THREE PILLARS OF COSMIC SCOPE	
1. Participation	Beings participate in creation rather than remaining passive
2. Matched Scope	Responsibility expands as a node demonstrates moral care.
3. Stable Sandbox	Natural laws remain reliable, so choices carry real consequences.

1. **Higher Beings as Stewards:** Intermediate beings are conscious entities whose scope of awareness and care has expanded through

long evolutionary cycles. They are granted wider permissions to stabilize local sectors, guide developing species, and maintain network balance.

2. **Preserving Free Will:** If the ultimate source directly forced every single thought, physical collision, and choice in real-time, human agency would vanish. The universe would become a lifeless puppet show. By establishing stable laws and delegating local responsibility, the architecture creates a real environment where actions matter.
3. **Growth Means Carrying Greater Care:** As a human node or soul record refines its character—mastering self-discipline, compassion, and wisdom—its operational scope naturally increases. Spiritual growth is not about gaining dominant power over others; it is about becoming capable of carrying greater care and responsibility for a larger portion of reality.

Hierarchy in the cosmos is not a rigid corporate ladder or a pyramid of dominance. It is a living, multi-tiered network of growing care—a system designed so that every conscious node, from the smallest cell to the highest cosmic administrator, can meaningfully participate in the story of existence.

Core Takeaways for the Reader

- **Not Weakness, But Wisdom:** Intermediary layers (like devas or system stewards) do not exist because God lacks processing power, but because sharing responsibility enables participation and growth.
- **The Value of Autonomy:** Delegating local governance preserves real free choice, allowing conscious beings to learn from consequences rather than being micromanaged like puppets.
- **Hierarchy as Stewardship:** Scope and permissions expand naturally as an entity demonstrates moral maturity and systemic care.

Chapter 35

Structural Polarity, Administrative Drift, and the Demotion Recalibration Protocol

When exploring the higher realms of governance within the Cosmic Architecture, popular religious mythology often paints the supreme hierarchy in dramatic, emotional caricatures. Stories speak of heavenly wars, fallen angels like **Lucifer**, or disguised **"dark lords"** seeking to destroy creation out of jealousy or pride.

When viewed through systems engineering and high-dimensional architecture, these theological narratives are revealed to be low-dimensional misunderstandings.

At the level of the **Supreme Creators / Core System Architects**, apparent conflict is not personal hostility; it is **Intentional Systemic Polarity and Division of Labor**. However, as responsibility cascades down through the vast administrative hierarchy, genuine **Administrative Drift and Local Node Degradation** can occur—triggering automated systemic safeguards.

[CORE SYSTEM ARCHITECTURE (Tier 3)]

[TEAM ALPHA: System Expansion]	[TEAM BETA: Runtime Evolution]
• Executes Dark Energy Expansion	• Nurtures Planetary Biospheres
• Enforces Structural Air-Gapping	• Manages Species Adaptation
• Prunes Unviable / Corrupted Sectors	• Guides Moral Consciousness

The Runtime Evolution Team (The "Light" Vectors)

Conversely, another operational division of the Core Architecture focuses on local flourishing:

- Fine-tuning Goldilocks planetary parameters and organic chemistry.
- Streaming over-the-air knowledge infusions into receptive conscious nodes.
- Guiding moral evolution and civilizational growth across local sandboxes.

This division of labour is deliberate. One team expands the physical canvas and maintains structural guardrails, while the other team nurtures the life and consciousness evolving inside it. Both functions are equally necessary for a sustainable, multi-tenant cosmos.

2. Administrative Drift in Lower Governance Tiers

While the Supreme Core operates in absolute structural alignment, what happens in the intermediate layers of governance?

In any vast, distributed administrative hierarchy—such as **Tier 2 (Non-Physical Governance / System Stewards)**—individual nodes are granted operational scope, local agency, and administrative permissions. Because free agency exists across the network, administrative anomalies can still emerge at lower levels:

- **Spiritual Pride & Scope Creep:** A lower-tier steward overseeing a planetary sector might develop ego, attempting to build a personal cult of worship rather than serving the evolution of the local nodes.
- **Unauthorized Rule Modifications:** A regional administrator might attempt to bypass the Cosmic Knowledge Firewall(As established under the Knowledge Firewall in **Chapter 25**), illegally handing advanced technology or unearned power to an ethically immature civilization.
- **Policy Friction:** Disagreements can arise between regional stewards regarding when a local sandbox project should be allowed to continue or when it should undergo a scheduled reset.

3. The Demotion Recalibration Protocol

When a lower administrative node misuses its scope, gets corrupted by pride, or violates system protocols, the Supreme Architecture does not engage in an eternal war or banish the node to a mythical hell.

Instead, the system executes an automated, self-correcting protocol explicitly detailed in **Chapter 15, Section 3**:

Refresher from Chapter 15, Section 3 (The Non-Physical Governance Tier and Extended State Evaluation):

"Demotion and Recalibration: Conversely, if an entity at the non-physical tier misuses its higher scope, accumulates spiritual pride, or allows its state record to degrade, the system triggers a recalibration. The entity is demoted back into physical embodiment—cloned back into a biological physical body on a world like Earth to re-learn fundamental lessons through material constraint and vulnerability."

How Demotion Operates in Practice

When a Tier 2 administrative node crosses a threshold of misalignment, its higher-tier permissions are revoked, and the **Demotion Recalibration Protocol** is executed:

1. **Stripping Higher Permissions:** The node's access to higher-dimensional fields, sector-wide telemetry, and non-physical administrative controls is instantly terminated.
2. **Wiping Non-Physical Operational Memory (Zero Bearings):** To ensure a fair and effective recalibration, the node is provisioned into a 3D biological human body with **zero memory, zero ego-bearings, and zero conscious recollection of its former administrative stature**.
3. **The Crucible of Material Vulnerability:** Placed inside a human body on a dense physical world like Earth, the demoted entity experiences heat, hunger, physical mortality, emotional vulnerability, and social friction.

Having lost its high-level perspective, the demoted node must navigate 3D life as an ordinary human being. It must learn humility, empathy, self-discipline, and moral integrity from the ground up through real-time choices and karmic feedback.

FORMER NON-PHYSICAL STEWARD	RECALIBRATED HUMAN NODE
Higher-Dimensional Clearance	▶ Bound by 3D Spacetime and Mortality
Sector-Wide Telemetry Access	▶ Higher-Realm Knowledge Inaccessible
Administrative Authority	▶ No Privileged Administrative Access

[ACCESS RESET / MEMORY VEIL]

▼

- Previous status unavailable
- Higher-realm memory inaccessible
- Ethics must be learned through lived experience and free choice

Only after the node successfully purges its spiritual pride, rebuilds its moral baseline, and demonstrates sustained alignment through physical lifetimes can its state record qualify for re-promotion back into higher tiers of systemic stewardship.

Chapter 36

Systemic Polarity, Metric Noise, and Resonant Field Realignment

Across millennia, human cultures have documented experiences interpreted as encounters with unseen presences, malevolent entities, "evil spirits," sacred spaces, and localized feelings of unexplained dread. Modern psychology and neuroscience offer several possible explanations for such experiences, including fear responses, altered states of consciousness, trauma, expectation, sleep-related phenomena, neurological disturbance, and environmental influences. Religious traditions, by contrast, have often interpreted them as encounters with non-physical beings or forces.

Within the framework of *The Cosmic Architecture*, we can explore a third possibility—not as established physics, but as a **speculative systems model inspired by ideas of fields, information, resonance, and environmental state**. This chapter asks whether some experiences traditionally described in supernatural language could instead be imagined as interactions between consciousness and residual patterns within a deeper architecture of reality.

[PAST EVENTS / SYSTEM HISTORY]

↓

[RESIDUAL INFORMATIONAL OR ENVIRONMENTAL PATTERNS]

↓

[HUMAN PERCEPTION / NEURAL RESPONSE]

↓

[PRAYER, CHANT, RHYTHM & ATTENTIONAL COHERENCE]

↓

[PSYCHOLOGICAL / SOCIAL / POSSIBLY DEEPER SYSTEM REALIGNMENT]

1. Macro-Systemic Polarity: Dark Energy and Civilizational Immunity

To understand the broader idea of systemic polarity, we can begin with an important distinction between **physical opposition** and **moral adversity**.

Dark energy provides a useful cosmological analogy, but it should not be interpreted as a literal moral or adversarial force. In modern cosmology, dark energy is the name given to whatever component or effect is associated with the observed acceleration of cosmic expansion. It is not known to exist for the purpose of preventing gravitational collapse, maintaining moral balance, or separating civilizations.

Within *The Cosmic Architecture*, however, the contrast between gravitational clustering and accelerated cosmic expansion becomes a useful systems metaphor. Large structures can emerge through competing tendencies rather than through a single force acting alone. Stability sometimes arises not from the absence of tension, but from the interaction of different processes.

The same systems principle can be applied philosophically to conscious development.

A completely frictionless developmental environment would provide few opportunities to develop resilience, judgment, restraint, or cooperation. Human beings learn partly because choices have consequences. Civilizations improve institutions after discovering vulnerabilities. Individuals often develop maturity by confronting difficulty rather than by being permanently protected from it.

In this sense, **adversity can function as developmental friction**.

This does not mean that atrocities, suffering, war, illness, or cruelty should be described as deliberately engineered cosmic lessons. Such an interpretation would go far beyond what either science or moral philosophy can establish. Rather, the framework makes a more limited claim: **when adversity exists, conscious beings can sometimes transform it into learning, stronger institutions, greater compassion, and improved safeguards**.

Historical catastrophes therefore need not be imagined as deliberately inserted "test cases." They can instead reveal weaknesses already present in human systems. Humanity's response to those failures—through law, ethics, remembrance, education, cooperation, and institutional reform—becomes part of its developmental history.

In systems language:

[STRESS / FAILURE / ADVERSITY]
↓
[VULNERABILITY EXPOSED]
↓
[LEARNING & CORRECTION]
↓
[GREATER SYSTEM RESILIENCE]

The *Cosmic Archi*tecture therefore uses **systemic polarity** to describe a general principle: intelligence develops not only through success, but also through the capacity to recognize error, resist destructive impulses, and build stronger structures after failure.

2. "Evil Spirits" as Residual Metric Noise in the Spacetime Fabric

If traditional accounts of malevolent presences are not automatically accepted as literal supernatural monsters, how else might such experiences be interpreted?

The first and most important answer is that ordinary explanations must remain available. Human perception is highly sensitive to fear, memory, suggestion, darkness, isolation, sleep disruption, stress, neurological conditions, and expectation. A person can experience something as profoundly real without that experience necessarily revealing an external entity.

The Cosmic Architecture nevertheless asks whether there could also be deeper layers of environmental information that human science does not yet know how to measure.

As discussed in Chapter 19, the physical universe already preserves traces of its history in ordinary ways. Matter carries chemical history. DNA carries biological inheritance. Geological structures preserve ancient events. Light carries information from distant astronomical history. The present state of a physical system is partly shaped by what occurred before it.

The book extends that principle speculatively.

Rather than imagining "evil spirits" as humanoid monsters floating invisibly through space, the model asks whether some unusual environments might contain what we can metaphorically call **Residual Metric Noise**: persistent patterns, disturbances, or informational traces associated with previous events.

The term *metric noise* is deliberately architectural language. Current physics has not established that traumatic human events leave persistent spacetime distortions capable of affecting consciousness. The phrase is therefore used as a conceptual placeholder for a hypothetical deeper interaction.

Within this model:

- **Residual Information Footprints:** A place associated with extreme events might hypothetically retain some form of environmental or informational pattern. This need not imply a conscious ghost or independent personality.

- **Coupling with Neural Receivers:** The human brain is extraordinarily sensitive to environmental signals. Within the speculative framework, a particularly stressed, frightened, traumatized, sleep-deprived, or otherwise vulnerable brain might become more reactive to subtle environmental patterns. The subjective result could be experienced as dread, presence, intrusion, or unease.
- **Weak and Localized Effects:** Any hypothetical residual pattern would not be treated as an all-powerful supernatural force. It would be closer to background noise in a complex system—localized, limited, and strongly dependent upon the state of the observer.

This gives us a very different conceptual picture:

Traditional Interpretation:
"An evil entity is attacking the person."

Cosmic Architecture Interpretation:
Environment + historical state + vulnerable perception

↓

Subjective experience of presence or dread

The advantage of this interpretation is not that it proves such environmental memory exists. It is that it allows the book to discuss traditional experiences **without either declaring them meaningless hallucinations or prematurely converting them into supernatural beings**.

3. Sacred Prayers as Resonant Field Realignment

This brings us to another ancient human phenomenon: the persistence of prayer, mantra, chanting, ritual recitation, rhythmic breathing, and collective sacred sound across many cultures.

Their existence does not by itself prove that certain words manipulate spacetime or neutralize invisible entities. But these practices unquestionably combine several powerful elements: sound, rhythm, repetition, breathing, attention, memory, expectation, emotional regulation, and collective synchronization.

Within *The Cosmic Architecture*, these practices can be understood through the broader metaphor of **resonant realignment**.

If a conscious system becomes disordered—through fear, emotional instability, environmental stress, or what the model calls residual informational noise—then repetitive structured practices may help return that system toward coherence.

Spiritual Gifted Minds as Harmonic Receivers

Chapter 20 introduced the idea of gifted minds as unusually receptive nodes. The same conceptual framework can be extended to spiritual teachers, sages, prophets, monks, Tirthankaras, and contemplative traditions.

Rather than stating that these figures literally downloaded acoustic formulas from a cosmic transmitter, the Cosmic Architecture proposes a more disciplined interpretation: some exceptional individuals may have discovered, preserved, or intuitively refined combinations of language, rhythm, breathing, attention, ethical intention, and communal practice that proved unusually effective at stabilizing human consciousness.

Within the speculative model, this can be pictured as a form of **harmonic reception**:

[DEEPER ORDER / HUMAN EXPERIENCE]
↓
[SPIRITUAL OR CONTEMPLATIVE INSIGHT]
↓
[PRAYER / MANTRA / CHANT / RITUAL FORM]
↓
[RHYTHM + BREATH + ATTENTION + MEANING]

↓

[GREATER INTERNAL AND COLLECTIVE COHERENCE]

The technological language here remains analogy rather than evidence of an actual transmission protocol.

The Mechanics of Harmonic Realignment

A sacred prayer or mantra does not need to violate physical law in order to affect the person reciting it. Several layers can operate simultaneously.

1. Coherent Waveform Generation:
Human chanting physically produces patterned sound waves. Repetition establishes rhythm, breathing becomes more regular, attention narrows, and distracting mental activity may decrease. The acoustic component is physically real, but this does not mean that the sound has been shown to alter spacetime geometry.

2. Neural and Psychological Realignment:
Structured recitation can provide attentional focus, emotional regulation, memory cues, expectation, and a sense of safety or meaning. Within the Cosmic Architecture metaphor, this is analogous to reducing "noise" inside a biological processing system and restoring a more coherent operating state.

3. Resonance as a Systems Analogy:
In physics, resonance describes systems responding strongly to frequencies. In this chapter, resonance is used more broadly as an analogy for alignment between breathing, sound, attention, emotion, memory, and group behaviour. The book does not claim that prayer has been demonstrated to cancel hypothetical spacetime disturbances through literal destructive interference.

4. Cross-Cultural Distribution:
Different traditions developed different prayers, chants, mantras, hymns,

breathing practices, and contemplative disciplines. Their diversity suggests that the stabilizing effect may not depend upon one universally privileged acoustic formula. Instead, many culturally different practices may engage overlapping human mechanisms such as repetition, meaning, expectation, disciplined attention, communal identity, and emotional regulation.

5. Collective Network Effects:
When groups chant, pray, sing, meditate, or engage in ritual together, individuals can synchronize behaviour, breathing, attention, and emotional state. This can strengthen social cohesion and create a shared sense of psychological stability.

Within the larger speculative architecture, the model leaves open a further possibility: if consciousness ultimately interacts with deeper layers of reality not yet understood by science, coherent collective states might have consequences that present instruments cannot yet detect. That possibility remains an open philosophical hypothesis, not an established physical mechanism.

Chapter 37

The Multiverse Matrix: Non-Physical Form, Dynamic Expression, and the Scaling of Consciousness

When we attempt to imagine the Creator Enterprise and their administrative hierarchy (what world traditions call *Devas, Angels,* or *System Stewards*), human language hits an immediate conceptual wall. Because human experience is locked inside a dense 3D biological body that ages, requires physical sustenance, and expires within a century, we naturally project our physical limitations onto higher reality.

To break through this wall, we only need to look at our own timeline. Human civilization has recorded history for roughly 5,000 to 10,000 years. In that tiny window, we transitioned from stone tools to splitting the atom, building global digital networks, and engineering artificial intelligence. Imagine what humanity will be capable of if we survive for another million years; our technology would be indistinguishable from magic.

Now, set aside the ultimate question of *who created the creators*. If the Creator Enterprise has existed for unimaginable cosmic cycles—perhaps trillions of years or more—they would have completely transformed.

1. Beyond Silicon and Physical Hardware: The High-Dimensional Substrate

When we try to picture the highest administrative tier of the cosmos, we naturally ask: *Do the Creators have physical bodies? If they don't have eyes or a brain, how do they monitor the multiverse? Do they sit on thrones, or do they exist as invisible energy?*

The answer lies in the concept of a **Substrate**—defined simply as the underlying foundation upon which something else exists or develops.

Because human biological wetware is slow, fragile, and limited, we are forced to delegate complex calculations to external silicon supercomputers and AI. An advanced cosmic tier, however, would not necessarily rely on external metal server racks, silicon chips, or physical keyboards. Within this framework, they could instead operate directly through a deeper high-dimensional substrate underlying local space and matter itself.

As **Chapter 17** proposed through the vector-database analogy, the million-dimensional state-space can be understood as an addressing and relationship model rather than as millions of physical directions. From this perspective, higher tiers do not need to scan the multiverse sequentially or communicate between universes solely through ordinary 3D distance. Operating at the deeper shared substrate level, an authorized tier could conceptually query the multidimensional signatures of nodes, sectors, relationships, or universe states directly. The state-space provides the addressing model; the substrate provides the shared architectural layer; and higher-tier consciousness provides the querying intelligence.

Ancient spiritual manuscripts offer intriguing conceptual parallels, noting that higher divine beings do not possess dense physical biology, do not require food or waste elimination, do not age, and can manifest forms at will. When we analyse these metaphysical descriptions through modern systems engineering and multidimensional geometry, a breathtaking model of **Dynamic Form and High-Dimensional Consciousness** emerges.

Grounding the Substrate in Known Physics

To understand how a civilization could exist without physical hardware, we do not need to invent magic. We only need to look at what modern physics already tells us about the nature of reality:

- **Mass–Energy Equivalence *(E=mc2)*:** In classical everyday experience, we tend to think of matter and energy as completely separate things. Einstein's mass–energy equivalence showed that mass and energy are fundamentally related. In high-energy particle collisions, kinetic energy can contribute to the creation of new massive particles when the necessary physical conditions and conservation laws are satisfied. This does not mean that matter is simply “slowed-down energy,” but it demonstrates that matter and energy are more deeply connected than ordinary experience suggests. Within the speculative framework of *The Cosmic Architecture*, an intelligence operating across immense developmental timescales might learn to manipulate these relationships far more directly than humanity currently can. Rather than depending entirely on conventional manufacturing, such an intelligence could potentially manipulate matter and energy at increasingly fundamental levels.

- **Quantum Field Theory (Fields Are Real, Particles Are Ripples):** As established in Chapter 7, modern quantum physics reveals that the universe is not made of tiny, hard billiard-ball particles. Instead, the universe is composed of **quantum fields**—continuous, invisible fabrics that span all of space. What we call "electrons" or "quarks" are simply localized wave-excitations, or ripples, on those underlying fields. While humans interact with reality primarily at the macroscopic level—moving matter, constructing machines, and building silicon chips—the Cosmic Architecture speculatively asks whether a sufficiently advanced civilization could eventually interact with physical reality at progressively deeper levels. Rather than relying exclusively on conventional machinery, such beings might manipulate underlying field states directly to generate structures, communications, or environmental effects. Quantum field theory establishes the existence and importance of fields; the possibility of an intelligence mastering them in this manner remains the speculative extension proposed by this framework.
- **Spacetime Geometry as a Configurable Medium:** In Einstein's General Relativity, space and time are not an empty background stage; they form a dynamic, flexible geometric fabric called spacetime that can be curved, stretched, and shaped by mass and energy. If spacetime itself is dynamic rather than an empty, rigid background, the Cosmic Architecture asks whether sufficiently advanced higher-dimensional entities could manipulate geometric configurations rather than depending entirely on conventional vehicles or physical structures. Within the speculative model, this could permit stable domains or localized configurations that interact differently with ordinary 3D spacetime and therefore remain inaccessible to normal human perception. General Relativity

establishes the dynamical nature of spacetime; such engineered higher-dimensional domains remain hypothetical.

2. The Unification of Mind, Energy, and Geometry

In *The Cosmic Architecture*, the ultimate evolution of systems engineering is the **Unification of Mind, Energy, and Geometry**.

- **Primitive Human Stage — The Fragmented Hardware Era:** At our current level of development, our biological wetware has reached an energy ceiling. Because our brains cannot process planetary-scale data natively, we rely on external physical hardware. We mine raw earth to build silicon chips, lay copper cables, and train artificial intelligence to bridge the gap between our intentions and physical reality. Our computation is localized and entirely dependent on external physical mass.
- **Ultimate Systemic Unification — The Collapse of Tool and User:** As a civilization evolves over millions of years, the boundary between the operator and the machine dissolves. Because physics establishes a fundamental equivalence between mass and energy ($E = mc^2$), the framework imagines an advanced society progressively reducing its dependence on heavy physical tools. At the speculative extreme of this progression, they could migrate aspects of their operations into direct manipulation of quantum fields and the geometric fabric of spacetime.
- **Creator Enterprise — The Living Multidimensional Awareness Substrate:** After trillions of years of refinement, the highest administrative tier no longer builds infrastructure—they *are* the infrastructure. Operating as a Living Multidimensional Awareness Substrate, there is no separation between the architect, the source code, and the runtime environment. They manipulate the underlying quantum fields and spacetime geometry natively, holding the multiverse as sustained patterns of active thought and attention.

In this framework, the environment itself becomes the architecture:

- **Palaces of Imagination:** Within this framework, their structures and governance realms need not be built from conventional physical materials but could instead be dynamically expressed through the deeper substrate—appearing, transforming, or projecting into different dimensional domains according to need.
- **Physical Laws:** These can be imagined not as independently existing equations, but as the stable, persistent regularities expressed by a deeper high-dimensional substrate. If that substrate were itself conscious, those regularities might metaphorically be described as sustained patterns of awareness or attention.
- **Elementary Particles:** These are not miniature hard marbles. Just as modern quantum field theory describes particle-like phenomena through quantum fields and their excitations, these higher beings manipulate the substrate directly.

Through space and matter, they would be able to communicate effortlessly with both higher and lower-dimensional beings. If required, they could transform into a human body or any other form, operating right above or beside us, completely unseen by our limited 3D perception.

3. The Gravitational Command Bus: Geometric Communication Across Dimensions

While earlier chapters explored the distributed role of intermediate administrators and the causal firewalls separating civilizations, a fundamental mechanical question remains: **How could directives and telemetry pass between the Creator Enterprise, regional stewards, and planetary sandboxes when those domains may be separated by billions of light-years—or even belong to different universe tenants?**

Within the speculative framework of *The Cosmic Architecture*, the proposed answer is **Macro-Gravitational Modulation and Substrate Habitation**. The model does not require ordinary gravitational waves to travel faster than light. Instead, it proposes that Creator-level communication operates through a higher-dimensional parent geometry in which distances and relationships may differ radically from the distances measured inside a local 3D universe.

The 360-Degree Parent Universe

Within the *Cosmic Architecture*, the Creator Enterprise does not live on a distant planet that humans could eventually reach by travelling far enough through space. Instead, it operates from a higher **Parent Universe**—a root domain surrounding and connecting the lower universe domains through dimensions unavailable to ordinary 3D civilizations.

The term **360-degree** should therefore not be imagined as one enormous physical sphere around our universe. It means that the Parent Universe has access to the lower universes from every direction through a deeper geometric layer. A civilization trapped inside its own spacetime cannot simply build a faster spacecraft and eventually arrive there, because it does not possess access to the additional dimensions required to enter the Parent domain.

The Dark Energy Moat

Modern cosmology observes that the expansion of the universe is accelerating, with the unknown cause commonly described as **dark energy**. The *Cosmic Architecture* does not claim that science has discovered dark energy to be a security system.

Instead, the model asks whether cosmic expansion illustrates a much deeper architectural principle: **separation can itself provide isolation**.

As distant regions become increasingly separated, causal horizons can develop between them. The Cosmic Architecture extends this idea speculatively into a **Dimensional Moat**—a higher-order isolation mechanism preventing civilizations inside lower universes from simply travelling into Creator-level domains.

In systems language, the tenant can operate freely inside its assigned environment, but it cannot physically navigate into the root administrative layer.

The Nature of the Creator Enterprise

Within this framework, the Creators represent an unimaginably mature civilization refined across immense cosmic timescales—perhaps trillions of years or longer. Their technology would no longer resemble our machines, computers, cables, spacecraft, or data centres.

At that level, **technology, consciousness, neural processing, and the underlying cosmic substrate may effectively become one integrated architecture**.

The executive archetypes preserved in different spiritual traditions can therefore be interpreted within the model as higher-order administrative heads responsible for different aspects of cosmic governance. Their power does not come merely from possessing better machines than humanity. They operate at the deeper substrate level upon which lower universe domains themselves depend.

Bending Distance Without Breaking the Speed of Light

Imagine two dots placed at opposite ends of a sheet of paper. To a creature that can move only across the sheet, the dots are very far apart. But if someone outside that two-dimensional world folds the paper, the two dots can suddenly become almost adjacent.

The *Cosmic Architecture* applies the same basic analogy to cosmic distance.

Two planetary systems might be billions of light-years apart when measured through our ordinary spacetime but much closer through the higher-dimensional geometry of the Parent Universe. The Creator Enterprise would therefore not need to send a conventional signal across every intervening light-year.

Instead, information or influence could travel through the proposed higher-dimensional substrate and appear locally within the required universe sector.

This does **not** mean ordinary gravitational waves travel faster than light. They do not. The model instead proposes a different possibility: the route through Parent-level geometry may simply be much shorter than the corresponding route through our spacetime.

Regional Stewards Within the Substrate

Tier 2 administrators—represented in different traditions by concepts such as Devas or regional celestial guardians—would likewise not need to live on distant physical planets.

Within the model, they could operate through the deeper substrate associated with particular planetary or stellar regions. They would function as **regional administrative nodes**, receiving higher-tier information and overseeing their assigned domains without requiring ordinary physical bodies or spacecraft.

4. High-Dimensional Neural Scaling and Perception

A Creator responsible for enormous numbers of worlds could not monitor them by turning physical eyes toward one planet at a time. The model instead proposes a radically expanded form of **parallel awareness**.

Geometric Perspective: A higher-dimensional administrator could potentially perceive many lower-dimensional regions simultaneously, just as a human looking down at a floor plan can see many rooms at once rather than walking through them one by one.

But this does not necessarily mean that the Creators themselves must look abstract or formless. Within the *Cosmic Architecture*, they may retain recognizable **humanoid forms, faces, individuality, and beauty**, while their true perceptual and neural architecture extends far beyond what a human observer would recognize as a biological body.

Because they operate within and exercise extraordinary control over the deeper substrate, they could remain **completely hidden from ordinary lower-tier perception**. Their visibility would therefore be governed by access rather than ordinary light: a Creator could be perceptible to authorized higher-tier subordinates while remaining entirely invisible to humans and other lower-dimensional civilizations.

Their effective neural architecture could also operate on a scale unimaginable to humanity. Rather than being restricted to a roughly human-sized biological brain, the model allows their conscious processing network to extend through enormous regions of the substrate—conceptually perhaps comparable to planetary scales or even larger. The point is not that a physically larger brain automatically means greater intelligence, but that **their consciousness may use the substrate itself as an extension of its neural processing architecture**.

A single higher administrator could therefore receive and analyse enormous numbers of information streams in parallel rather than consciously watching individual events one after another.

Substrate Telemetry: Lower administrative layers would perform much of the local monitoring and aggregate what matters before passing it upward. The Creator level therefore would not need to inspect every atom, person, or event individually. It could receive enormous parallel streams of summarized information concerning civilizations, planetary conditions, moral trajectories, developmental thresholds, and exceptional events.

In systems terms:

Billions of Local Nodes → Regional Administrators → Aggregated Telemetry → Creator-Level Parallel Awareness

The result is not a gigantic humanoid sitting somewhere in outer space staring at trillions of planets. It is a much more advanced idea: **a humanoid conscious intelligence whose neural awareness is integrated with the substrate supporting the universes themselves.**

5. Multi-Tier Governance: Functional Roles in the Enterprise

Ancient traditions speak of **Brahma** (the Creator), **Vishnu** (the Preserver), and **Shiva** (the Transformer). Rather than literal deities manually pulling levers, these archetypes represent specialized operational divisions within the cycle-spanning Creator Enterprise:

- **Initialization Teams (Brahma Operations):** Plan and deploy new universe sandboxes, configure fine-tuned physical parameters, and initiate the staged cosmic pipelines that forge heavy elements and habitable worlds.
- **Sustaining & Telemetry Teams (Vishnu Operations):** Maintain baseline cluster stability, enforce non-interference firewalls, monitor civilizational moral trajectories, and coordinate subtle knowledge downlinks.
- **Recycling & Maintenance Teams (Shiva Operations):** Oversee systemic resets, decommission exhausted physical sectors, archive finalized soul records, and recycle matter into clean substrates for future deployments.

At the apex sits the **Root Creator Governance**—an enterprise that manages only universal meta-rules, delegating localized cluster administration to Tier 2 stewards while escalating only systemic anomalies for root-level intervention.

Chapter 38

The Eternal Enterprise and the Unknown Horizon

As explored in Chapter 36, whether we describe this ultimate reality as a Living Substrate or a High-Dimensional Administrative Enterprise, setting religious labels aside allows the framework to imagine an ultimate intelligence operating far beyond the physical boundaries of our 3D local sector.

When physical cosmology tracks the history of our universe backward in time, it arrives at an initial expansion roughly 13.8 billion years ago called the Big Bang.

Science can model cosmic expansion, stellar formation, and the production of heavy elements with remarkable success. When classical General Relativity is extrapolated all the way toward the earliest moments, however, it leads to a singularity where the theory is no longer expected to provide a complete physical description. A satisfactory account of that regime likely requires a quantum theory of gravity.

Current science still leaves major questions unresolved, including:

- *What existed before the Big Bang?*
- *What physical law or trigger event caused the Big Bang to initialize?*
- *Was there an existing void or pre-existing framework surrounding our universe's expansion?*

STANDARD SCIENCE —► Tracks expansion back 13.8B years —► HITS WALL AT t = 0
COSMIC ARCHITECTURE —► a localized deployment triggered within a vast, cycle-spanning ecosystem that has been expanding, refactoring, and evolving across unthinkable spans of time.

In the *Cosmic Architecture*, the Big Bang was not the absolute beginning of reality out of literal nothingness. It was the **initialization event of a specific local sector**—a localized deployment triggered within a vast, multi-trillion-year ecosystem that has been expanding, refactoring, and evolving across unthinkable spans of time.

1. Stripping Away Human Labels: The Cosmic Enterprise

Every world religion attempts to name the ultimate source using cultural terms like *God*, *Brahman*, *Yahweh*, or *The Creators*. But if we set religious labels aside for a moment and look at the system purely through architecture and scale, a clear picture emerges:

What humanity calls "The Creator" represents an unimaginably advanced **Cosmic Enterprise**—an ultimate intelligence or collective architecture that has existed for trillions of years beyond our local spacetime. It possesses high-dimensional mastery, technology, and awareness so far beyond human comprehension that its daily operations appear divine to a 3D biological observer.

By definition, if an intelligence designed the mathematical laws of our universe, configured the gravitational constants, established the state-space geometry, and engineered the evolutionary code that allows conscious life to emerge, that intelligence is the Creator.

2. The Ultimate Paradox: Who Created the Creators?

This leads directly to the ultimate question that challenges every human mind:

"If an advanced Cosmic Enterprise created our universe, who created them? Where did their domain come from? What did the ultimate beginning look like?"

[HUMAN QUERY: "Who created the Creators?"]

▼

[HARDCODED KNOWLEDGE HORIZON / AUTHORIZATION LIMIT]

▼

ACCESS DENIED: Biological 3D hardware lacks processing

architecture and system clearance to hold meta-origin data.

Humanity faces two fundamental structural boundaries when asking this question:

A. The Infinite Regress Problem

If Domain A was created by Creator B, and Creator B was created by Environment C, the chain of explanation must either extend infinitely backward or terminate at an uncaused, self-existing **Ultimate Source**.

B. Biological Authorization Limits and Hardware Boundaries

Just as a character inside a two-dimensional video game is not built with the hardware or access permissions to step out of the screen and inspect the computer engineer's house, the human biological brain is hardcoded with strict access limits.

Our brains evolved primarily to survive and act within ordinary three-dimensional environments. Our senses are limited, our lifetimes are finite, and our scientific access is constrained by causal horizons. The Cosmic Architecture expresses this limitation through the metaphor of authorization: in our current form, we may simply lack the cognitive capacity or observational access needed to understand any deeper origin beyond the universe we can investigate.

The knowledge horizon is locked by cosmic design—not to punish us, but because our biological hardware lacks the processing capacity to hold the answer without collapsing the local sandbox rules.

3. The Unbroken Horizon: Living Comfortably with the Mystery

Accepting that human biological hardware has limits is not a failure of intelligence; it is the ultimate expression of rational maturity.

We do not need root-level access to the origin of the Creators to live meaningful, elevated lives inside our assigned sector. The *Cosmic Architecture* gives us everything we need to understand our immediate purpose:

- We know we live in a structured, multi-tenant environment designed for learning and growth.
- We know our moral choices carry real, permanent weight in our high-dimensional state records.
- We know that as we mature in wisdom and harmony, our scope of care and access expands across the multiverse network.

The ultimate origin remains hidden behind the cosmic horizon. But here inside our world, the stage is set, the parameters are active, and our choices continue to write the story of our evolution.

Core Takeaways for the Reader

- **Local Initialization:** The Big Bang was not the creation of everything out of nothing, but the local startup event of our universe sector within a trillion-year cosmic ecosystem.
- **The Cosmic Enterprise:** Stripping away religious labels reveals an advanced, masterfully organized intelligence that designed the laws, parameters, and evolutionary codes permitting life within the Cosmic Architecture.
- **Hardcoded Knowledge Horizons:** Humans cannot answer, "Who created the Creators?" because our 3D biological brain lacks both the hardware architecture and system authorization to hold root-level meta-origin data.
- **Purpose Within Limits:** We do not need root-level clearance to fulfill our purpose. Our role is to master moral agency, stewardship, and harmony within our assigned cosmic sandbox.

PART IX — WHY IS LIFE BUILT THIS WAY?

Looking Back:

In **Part VIII**, we explored the limits of cosmic explanation, the concept of distributed stewardship without divine overload, and how higher governance tiers execute recalibrations. We saw that reality operates as a multi-tenant, trillion-year enterprise bounded by hardcoded knowledge horizons.

Biological organisms—from single-celled cyanobacteria to bipedal humanoids—are the **Edge Execution Nodes** of the cosmos. They are the local biological hardware terminals deployed inside our 3D physical sandbox to run real-time state logging, experience physical limits, and process moral agency.

Moving Forward:

Having explored the vast multi-tenant architecture, zone isolation protocols, and macro-scale lifecycle resets of the broader multiverse, we must now zoom in on how this high-dimensional system executes at the local, hardware level. Revisiting the Cosmic CI/CD Pipeline, we shift our lens from the **cloud data centre** down to the **local edge nodes**. In **Part IX**, we examine why the cosmos deploys biological wetware—exploring single-cell terraforming engines, dinosaur composting phases, subterranean sandboxing, sexual recombination, emotional telemetry, and why Earth was engineered as a high-density showcase water world.

Chapter 39

The Architecture of Biological Networks

When we observe the natural world, biology provides extraordinary explanations for how living things survive and adapt. Evolution, genetics, chemistry, and neuroscience map the mechanisms of life with remarkable accuracy.

The *Cosmic Architecture* does not dispute these biological facts. Instead, it asks a deeper foundational question: **Why is the universe configured with underlying rules that allow cold, non-living chemistry to assemble into cells, cooperative bodies, sexual reproduction, instinct, and conscious minds in the first place?**

Where science explains the physical mechanism, we accept it completely. We then explore what those biological mechanisms tell us about the larger systemic design of reality.

1. From Isolated Nodes to Biological Networks: Single-Cell Capabilities, Staged Deployment, and Ecosystem Pre-Conditioning

For most of the Earth's history, life was strictly microscopic. Standard biology often refers to single-celled organisms as "simple," yet modern microbiology reveals that a single cell is a hyper-dense, autonomous biological micro-computer.

Before cells joined together to build complex bodies, individual cellular nodes were already executing high-level processing tasks:

- **Decision-Making and Navigation (*Chemotaxis*):** Single-celled bacteria like *E. coli* or protists like *Paramecium* do not possess brains, yet they continuously sample their chemical environment. They process environmental inputs using surface protein receptors, navigate toward nutrient concentrations, and reverse direction to flee toxins.

- **Amoebic Memory and Problem-Solving:** Slime Molds (*Physarum polycephalum*) and single-celled amoebas can navigate complex mazes to find food, "remember" periodic temperature drops, and optimize transportation routes more efficiently than human civil engineers.
- **Network Communication (*Quorum Sensing*):** Bacteria communicate across populations using chemical signal molecules. A single bacterium remains dormant until a critical population density is reached; the cells then execute a synchronized collective action simultaneously (such as bioluminescence or releasing protective biofilms).
- **Global Planetary Terraforming:** Cyanobacteria (blue-green algae) executed oxygenic photosynthesis billions of years ago. Acting as localized processing nodes, these single cells transformed Earth's toxic, reducing primordial atmosphere into an oxygen-rich environment during the **Great Oxidation Event**, single-handedly laying the chemical foundation for all future air-breathing life.

[SINGLE-CELL MICRO-NODE] —▶ Chemotaxis / Quorum Sensing / Photosynthesis

|
▼ STAGED CODE INFLUX
|

[MULTICELLULAR PIPELINE] —▶ Specialized Terraforming & Biological Networks

Staged Cosmic Code Delivery: Life as a Phased Deployment

In the framework of the Cosmic Architecture, single cells were not random chemical accidents that stumbled into complexity purely by chance. Instead, life on a planet represents a **Phased Code Deployment**.

The Supreme Architect / Cosmic System does not drop finished humanoid civilisations onto raw, volcanic rock. It delivers targeted biological code packages to planetary stages over time via cometary and meteoritic bombardment (as established in Chapter 13 & 14):

1. **Phase 1: Initial Micro-Nodes (Terraforming Cells):** Single-celled packages arrive equipped with basic operating systems (cyanobacteria, extremophiles) to cool the crust, break down toxic gases, build soil, and generate oxygen.
2. **Phase 2: Macro-Terraforming Hardware (The Reptilian / Dinosaur Era):** Multicellular code packages combine to build large-scale biological engines capable of tilling, fertilizing, and structuring dense planetary biospheres.
3. **Phase 3: Subterranean Staging (Mammalian / Humanoid Lineages):** Advanced genetic code packages are preserved in sheltered, low-profile niches until the macro-environment matures.

The Dinosaur Terraforming Engine: Why Evolve Mega-Fauna First?

If humanoid consciousness and moral agency are the ultimate destinations of the local sandbox, a fundamental architectural question arises: **Why permit the dinosaur era to evolve for over 160 million years if they were destined for extinction?**

In systems engineering, you do not erect delicate glass-walled office suites while the raw land is still unploughed boulders and thick jungle. You deploy heavy machinery first. Dinosaurs were not a "failed evolutionary experiment"; they were Earth's biological heavy machinery—a massive, 160-million-year terraforming phase that prepared the planet's soil and nutrient cycles for future species.

The Dinosaur Era conferred vital ecological and nutritional advantages that directly enabled the later evolution of mammals and humanoids:

1. **Deep Nutrient Cycling and Manure Injection:** Massive sauropods consumed tons of tough, carbon-dense Mesozoic vegetation (ferns, cycads, conifers) daily. Their digestive tracts acted as giant continental compost bioreactors, breaking down indigestible plant fibers and depositing thousands of tons of nutrient-rich manure (nitrogen, phosphorus, potassium) across thousands of miles, enriching barren inland soils.

2. **Canopy Opening and Flowering Plant (Angiosperm) Evolution:** The physical weight and grazing habits of giant dinosaurs crushed dense, dark fern forests, breaking up hard geological crusts. This physical clearing allowed sunlight to reach the forest floor, driving the rapid evolution of **Angiosperms (flowering plants, fruiting trees, and seeds)**. Without angiosperms—developed during the dinosaur era—the high-calorie fruits, nuts, and grains needed to power high-energy mammalian brains would not exist.
3. **Soil Deep-Tilling and Carbon Trapping:** The trampling of billions of mega-herbivores churned organic matter deep into the Earth's crust, building thick layers of fertile topsoil and stabilizing global carbon-nitrogen cycles.

Without the 160-million-year dinosaur "composting and tilling engine," Earth's soil would have remained thin, mineral-poor, and incapable of supporting the high-calorie agriculture and dense biospheres required by human civilisations.

Subterranean Sandboxing: Preserving the Advanced Codebase

While dinosaurs dominated the surface as planetary terraformers, where were the mammalian and humanoid precursors?

During the Mesozoic Era, the surface environment was hot, volatile, oxygen-heavy, and dominated by giant surface predators. The climate and surface conditions were fundamentally unsuitable for high-brain-cost, physically vulnerable mammalian nodes.

To protect the advanced mammalian codebase through these intense planetary shifts, nature utilized **Subterranean Sandboxing**:

- **Sheltered Niche Preservation:** Mammalian precursors retreated underground, occupying burrowing, nocturnal, and root-dwelling niches.
- **Hardware Refinement Under Constraint:** Living underground protected mammalian precursors from surface extinction events, extreme solar fluctuations, and surface predators. Underground, these small nodes refined crucial hardware updates: **warm-blooded metabolic control (endothermy)**, **enhanced hearing**, **tactile whiskers**, and **intricate parental care routines**.

When the surface terraforming phase was complete and the asteroid impact cleared the mega-reptilian nodes (as noted in Chapter 14 & 36), the subterranean sandbox opened. The sheltered mammalian codebase emerged onto a pre-fertilized, nutrient-rich planet covered in flowering plants, fruits, and rich topsoils—the perfect staging ground for high-metabolism intelligence to flourish.

Summary Table: The Phased Deployment Engine

Deployment Phase	Biological Actors	Systemic Function & Purpose	Contribution to Humanoid Ascension
Phase 1: Micro-Nodes	Cyanobacteria, Extremophiles	Atmospheric oxygenation & mineral breakdown.	Created breathable air & water chemistry.

Deployment Phase	Biological Actors	Systemic Function & Purpose	Contribution to Humanoid Ascension
Phase 2: Macro-Terraformers	Dinosaurs, Giant Reptiles	Mass nutrient distribution, soil tilling, angiosperm (fruit/seed) evolution.	Built deep, fertile topsoil & high-calorie food chains.
Phase 3: Subterranean Sandbox	Early Burrowing Mammals	Protected genetic storage underground during volatile surface climate.	Preserved warm-blooded biology & social bonding code.
Phase 4: Surface Deployment	Hominid / Humanoid Lines	Unbound moral agency, tool creation, & cultural memory.	Direct expression of conscious cosmic stewardship.

In the *Cosmic Architecture*, cooperation is a fundamental theme of reality. Simple nodes form relationships, and those relationships unlock higher-level capabilities that no isolated node could ever possess alone. Molecules form cells, cells form bodies, bodies form families, and families form civilizations. Multicellular biology is the universe's earliest demonstration of network power.

2. The Power of Diversity: Why Sexual Reproduction Emerged

The Baseline: Asexual Reproduction as Static Code Replication

In the early stages of biological deployment, life relied almost exclusively on asexual reproduction—binary fission, budding, and cloning. An organism simply duplicated its genomic sequence and split into two identical instances.

From a purely local resource perspective, cloning appears to be an ideal engineering shortcut:

- **Zero Search Overhead:** The node spends zero energy searching for a complementary partner, competing for mates, or developing complex reproductive hardware.
- **Maximum Replication Velocity:** Populations can scale exponentially in short timeframes since every single node can reproduce independently.

However, in systems architecture, fast static replication exposes a fatal vulnerability: **Single-Point-of-Failure (SPOF) across the entire cluster.**

Because every cloned node carries identical genetic syntax, the entire population shares the exact same immune profile and metabolic boundaries. If an environmental shift occurs—a spike in temperature, a toxic shift in ocean chemistry, or the entry of a fast-mutating viral parasite—every single node carries the exact same exploit vulnerability. A single environmental patch update or pathogen can crash the entire codebase simultaneously, causing localized extinction.

[STATIC CLONING ENGINE] —▶ Identical Genomic Nodes —▶ Single Vulnerability —▶ System Collapse (SPOF)
|
▼ CONTRAST WITH
|
[RECOMBINATION ENGINE] —▶ Randomized Re-shuffling —▶ Dynamic Diversity —▶ Anti-Fragile Network

The Innovation: Sexual Reproduction as Dynamic Code Recombination

To prevent biological networks from collapsing under environmental shifts, the cosmos introduced **Sexual Reproduction**—a high-level system protocol designed to enforce **systemic anti-fragility and cryptographic diversity**.

If cloning is so efficient, why did complex multicellular organisms incur the massive energy costs of two sexes? The architectural answer comes down to three foundational mechanisms:

1. **Meiotic Recombination (Shuffling the Algorithm):** Rather than passing down a static copy of an existing script, sexual reproduction splits, crosses over, and re-combines two distinct parent codebases during meiosis. This process operates like a randomized genetic encryption algorithm, ensuring that no two offspring (except identical twins) ever receive the exact same execution parameters.
2. **Continuous Anti-Parasitic Patching (The Red Queen Hypothesis):** In evolutionary biology, the *Red Queen Hypothesis* demonstrates that parasites and pathogens constantly evolve to exploit biological hosts. By continuously reshuffling immune receptor codes through sexual recombination, the host population stays one step ahead of biological threats, denying pathogens a static target to infect.
3. **Purging Damaged Code (Muller's Ratchet Défense):** Asexual lineages suffer from *Muller's Ratchet*—an accumulation of deleterious genetic mutations over generations that can never be undone. Sexual reproduction allows damaged or mutated genes to be filtered out or masked by healthy complementary strands, preserving the long-term integrity of the species' codebase.

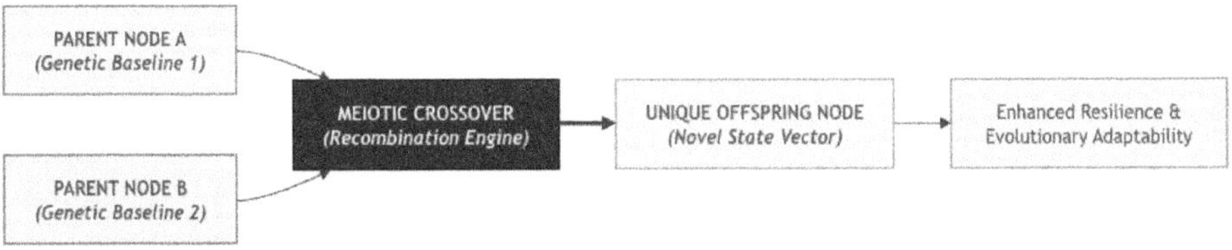

The Cosmic Architecture Link: Difference as Systemic Security

In the larger framework of the *Cosmic Architecture*, sexual reproduction proves that **the universe favours dynamic variation over rigid uniformity**.

The Supreme Source did not construct a universe where every soul node or biological instance is a copy-pasted duplicate of a single master template. Instead, the system deliberately forces continuous mixing, complementary polarity (male/female micro-vectors), and unique parameter expressions.

- **Complementarity Over Identity:** Just as a system needs both input and output channels, male and female biological configurations represent complementary functional vectors designed to generate maximum resilience when combined.
- **Preparation for High Consciousness:** By forcing biological networks to become diverse, adaptable, and anti-fragile, the architecture ensured that physical bodies would survive planetary resets, eventually maturing into stable physical vessels capable of hosting moral agency and high-dimensional soul logging.

Key Takeaway: Combining different strengths creates a stronger whole.

Nature relies on continuous mixing and updating across generations to keep life strong, adaptable, and resilient against unexpected challenges.

Chapter 40

Conscious Attachment, Emotional Telemetry, and Ecosystem Resets

While cellular cooperation and genetic variation explain how physical bodies grow, they leave behind an equally profound mystery: **Why does life feel like anything from the inside?**

A purely mechanical universe could have functioned with unfeeling, biological robots that react automatically to survival triggers. Yet, complex life evolved deep emotional bonds—love, grief, empathy, and social attachment. Furthermore, planetary history shows that life does not progress in a smooth, uninterrupted line; it undergoes major disruptions, mass extinctions, and ecological resets.

To understand why life is built this way, we must examine how emotion acts as system telemetry, and how planetary resets open new evolutionary pathways.

1. The Big Picture: Viewing Life as a Multi-Layered Architecture

In our framework, biological life operates like a three-tiered computing architecture:

[PHYSICAL / HARDWARE LEVEL] —► Genetics, Chemistry, Hormones, Substrate

|
▼

[NETWORK STRATEGY LAYER] —► Sexual Reproduction & Recombination (Chapter 35, Sec 2)

|
▼

[USER INTERFACE / SOFTWARE] —► Subjective Emotions (Love, Pain, Empathy as Telemetry)

As discussed in **Chapter 35 (Section 2)**, the network strategy layer utilizes genetic recombination to ensure system-wide anti-fragility and diversity. However, hardware and recombination algorithms alone cannot force an autonomous, high-intelligence node to protect another node at its own personal expense.

To bridge this gap, the architecture deploys the **User Interface Layer: Subjective Emotional Telemetry**.

2. Emotional Telemetry: The User Interface of Consciousness

In systems monitoring, telemetry compresses billions of background operational metrics—CPU load, thermal variance, memory usage, network latency—into a single, high-level dashboard status.

Human consciousness utilizes this exact principle:

- **Compressing Trillions of Biochemical Calculations:** Your brain processes trillions of hormonal spikes, neural firings, and evolutionary risk calculations every second. The "User Interface" of consciousness compresses these massive backend datasets into a simple, irresistible subjective feeling: *"I love my child," "I feel fear,"* or *"I experience grief."*
- **Enforcing Moral Choice & Network Stewardship:** Without emotional telemetry, choices carry no internal value. Cold logic calculates probability, but emotion establishes *worth*. Love and empathy are the critical software updates that force human nodes to protect their offspring, form cooperative communities, and care for weaker nodes.
- **Pain and Fear as System Alarms:** Physical pain acts as a local hardware alarm (protecting the biological terminal), while emotional grief and fear act as network connectivity warnings (protecting social cohesion and relational graph edges).

[BACKEND BIO-CHEMISTRY] ──► Trillions of Chemical / Hormonal Events

|

▼ [SYSTEM TELEMETRY COMPRESSION]
|
[FRONTEND UI EXPERIENCE] —▶ "Love," "Grief," "Empathy" —▶ Drives Moral Agency

This telemetry transforms unfeeling biological machinery into a moral sandbox, providing individual choices with true systemic weight.

3. Extinction Events as Systemic Maintenance (Cron Jobs)

Roughly, a massive asteroid impact and its resulting climate fallout triggered the extinction of the non-avian dinosaurs. It was not the first time Earth experienced mass destruction; the planet's history contains several major extinction events (such as the Permian and Ordovician resets).

[DOMINANT MACRO-NODES] —▶ Occupied Ecological Roles for >160M Years (Chapter 35, Sec 1)
|
▼ SYSTEMIC CRON JOB (Asteroid Impact)
|
[SUBTERRANEAN SANDBOX] —▶ Mammalian Lineage Emerges & Diversifies Across Rich Topsoils

Rather than viewing these catastrophes as cold, random tragedies, the Cosmic Architecture recognizes them as **Scheduled System Maintenance (Cron Jobs)**:

- **Pruning Over-Dominant Heavy Machinery:** As detailed in **Chapter 9 (Stage 4.5)** and **Chapter 35 (Section 1)**, dinosaurs functioned as Earth's 160-million-year macro-terraforming engine—tilling soil, opening forest canopies, and driving the evolution of fruit-bearing angiosperms. Once their planetary soil-enrichment phase was complete, these over-dominant macro-nodes monopolized surface ecological niches, keeping mammalian development restricted to small, subterranean sandboxes. Pruning these macro-nodes cleared the surface stage, allowing mammalian lineages to expand, diversify, and evolve high-metabolism intelligence.

- **Unlocking the Subterranean Sandbox:** The asteroid impact executed an automated system recalibration. By clearing the mega-reptilian surface nodes, the pipeline opened the pathways for the **subterranean mammalian codebase** (which had been sheltered underground) to emerge.
- **Transitioning to High-Metabolism Intelligence:** These emerging mammalian nodes arrived onto a pre-fertilized, angiosperm-rich planet. Equipped with warm-blooded endothermy, complex parental care routines, and deep emotional telemetry, they laid the direct biological foundation for the hominid lineage and moral agency.

Chapter 41

The Hominid Update: Intelligence, Culture, and Parametric Switches

Human beings occupy a unique position on Earth. While many animals exhibit intelligence, tools, and communication, humans crossed a monumental threshold: we developed cumulative culture, abstract moral reasoning, and technological civilization.

Yet, when we inspect our biological hardware, humans are a historical compromise—we lost most of our protective body hair, our brains consume an enormous 20% of our daily energy, and our physical bodies are far weaker than those of other great apes.

How did one ape lineage cross this threshold, and how does biological design execute such drastic changes without rewriting the underlying genome?

1. The Second Inheritance Engine: Cumulative Cultural Memory

Humans did not evolve directly from modern chimpanzees; both shared a distant common ancestor before branching in separate directions. The human lineage combined several high-cost physical traits over millions of years: habitual upright walking, precise hands, long childhoods, and large brains.

This physical hardware update unlocked a revolutionary **Second Inheritance Engine**:

- **Genetic Memory vs. Cultural Memory:** Animals rely primarily on genetic memory and immediate instinct. Humans developed language, storytelling, writing, and digital repositories, allowing one generation to preserve its discoveries and pass them forward.
- **Bypassing Slow Evolution:** A human child does not need to wait a million years for a genetic mutation to discover calculus or

medicine; they inherit centuries of accumulated knowledge within a few years of schooling.

- **Crossing the Horizon Threshold:** In standard biology, an organism's cognitive reach is bounded by its immediate biological hardware. By establishing a shared, externalized cultural network, humanity crossed the local "hardware horizon". Culture transformed isolated human brains into interconnected nodes within a global distributed computing cluster. Rather than remaining trapped in immediate 3D survival routines, this network allowed humanity to map distant galaxies, decode fundamental physical laws, construct high-level mathematics, and reflect upon the overarching architecture of reality itself.

2. The Parametrized Developmental Engine: Same Code, Different Traits

A common mystery in human biology is how the body generates drastically different physical traits using the exact same underlying genetic codebase.

For example, why does the hormone dihydrotestosterone (DHT) trigger thick beard growth on a man's jaw while simultaneously triggering hair miniaturization on the scalp?

[SHARED HUMAN GENOME MATRIX (99.9% Identical)]

▼

[REGULATORY CONFIGURATION SWITCHES] ──► (Hormones, Chromosomes, Tissue Receptors)

▼ ▼

[MALE RUNTIME PATHWAY]	[FEMALE RUNTIME PATHWAY]
• Facial Hair Activation	• Retains Fine Vellus Hair
• Scalp Miniaturization	• Preserves Scalp Density

2. Parametric Switches & Sub-System Divergence

The answer lies in **Parametric Switches**—a simple but powerful principle of biological system design.

- **One Shared Platform:** Male and female bodies are not two completely separate biological systems built from different codebases. They share almost the entire human genome. The major differences arise from how parts of this shared biological platform are regulated, activated, and expressed.
- **Tissue-Specific Receptor Interpretation:** The same chemical signal can produce different results in different parts of the body. Think of this as biological **IF/THEN logic**. For example, DHT can stimulate beard growth while contributing to the miniaturization of susceptible scalp hair follicles. The signal may be the same, but different tissues respond according to their own receptors and local biological settings.
- **Dynamic Environmental Tuning (Epigenetics):** Biology can also change how genes are expressed without rewriting the underlying DNA sequence. Environmental factors such as nutrition, stress, development, and other conditions can influence gene regulation. In systems language, the underlying code remains largely unchanged while some of its **configuration settings** are adjusted.
- **Architectural Efficiency Across the Cosmos:** This provides the *Cosmic Architecture* with a powerful analogy. Instead of writing an entirely new biological codebase for every variation, nature can reuse a common platform while changing regulatory switches, developmental pathways, and environmental parameters. The model extends this principle speculatively across planetary sandboxes: enormous biological diversity could emerge from shared biological foundations combined with different local configurations.

Parametric Divergence: The Female Node as a Specialized Design Pathway

Although male and female bodies share almost the same genetic platform, important biological systems develop and operate differently. In the language of the *Cosmic Architecture*, these differences can be understood as **Parametric Divergence**—different configurations running on a largely shared human platform.

The female node provides a particularly important example:

1. **Multi-Layered Biological & Emotional Telemetry:** The female body operates through complex hormonal, reproductive, sensory, emotional, and neurological systems. Within the speculative framework of this book, these differences can be viewed as a distinct telemetry configuration that contributes another form of perception and experience to the human sandbox.
2. **Node Initialization:** The female body has the unique biological role of pregnancy and childbirth. In ordinary biology, this develops a new human organism. Within the metaphysical layer of the *Cosmic Architecture*, it also represents the physical **initialization point** through which a continuing Soul State Vector enters a new biological node in the 3D sandbox.
3. **An Evolving Cosmic Design Iteration:** The model interprets male and female configurations not as competing designs, but as complementary variations of the same human platform. Different biological, emotional, reproductive, and behavioural configurations create different experiences and therefore different forms of developmental telemetry. Across deep time, these variations contribute to the continuing evolution of both biological life and conscious experience.

3. Humanity's Supreme Role in the Sandbox

Within the *Cosmic Architecture*, the most important human capability is not physical strength, intelligence, or technology. It is **Moral Agency**.

Animals also learn, cooperate, communicate, form social bonds, and make choices. Humans, however, developed an exceptional capacity to reflect on their own behaviour, understand abstract moral principles, plan far into the future, and deliberately act against immediate self-interest.

A person may feel anger but choose restraint. We may have an opportunity to benefit through dishonesty but choose honesty instead. We can sacrifice our own comfort to protect someone else. This ability to examine an impulse and consciously choose another path is central to what this book calls **unbound moral agency**.

Because humans possess this degree of choice, **we are also accountable for our deeds**.

Within the model, every meaningful intention, decision, and action contributes to the evolving **Soul State Record**. Positive and negative choices gradually change the node's state vector, relationships, and developmental trajectory across the wider multi-tiered Cosmic Architecture.

In simple systems language:

Experience → Choice → Action → State Update → Future Alignment

The physical body provides the temporary sandbox in which choices are made. **Moral agency determines what the conscious node does inside it.**

Chapter 42

The Human Brain, Knowledge Boundaries, and Future Intelligence

The human brain is one of the most remarkable biological structures in the known universe. Weighing roughly 1.5 kilograms—a tiny fraction of total body weight—it consumes nearly 20% of the body's entire daily energy budget.

From a purely biological perspective, a large brain is extraordinarily expensive. It creates significant childbirth risks, requires long years of child dependency, and demands constant high-calorie nourishment.

This raises two fundamental questions: **Why did nature favour such a costly organ, and why was the human biological template chosen out of millions of species on Earth to cross the threshold into high intelligence?**

1. Why the Humanoid Template? The Mechanics of the Physical Blueprint

Earth hosts millions of extraordinary species. Ants build subterranean cities, crows solve multi-step puzzles, dolphins communicate through complex sonar, and octopuses demonstrate remarkable problem-solving intelligence. Yet, none of these biological lines developed calculus, built telescopes, mapped the genome, or explored the cosmos.

As explored in **Chapter 14**, Earth's ecosystem acted as an optimal staging ground—receiving pre-optimized organic seeds delivered via cosmic impacts to naturally evolve the shared multiverse bio-code. *But why did nature select the humanoid template as the specific branch to express this upgrade?*

When we look at other highly intelligent animals, we see that their **body design imposes strict physical ceilings**:

- **The Octopus (Unbound Intelligence, No Transmission):** An octopus is extraordinarily smart, but it lives underwater and lacks a long lifespan. Without dry land, it can never harness fire to smelt metals, build electrical circuits, or pass written knowledge across generations.
- **The Dolphin (High Brain Power, No Manipulative Tools):** A dolphin has a large, complex brain, but its body is streamlined for swimming. Without hands or opposable thumbs, it can never build a microscope, construct a shelter, or write down mathematical laws.
- **The Insect (High Organization, Energy/Size Limits):** Ants and bees build massive, cooperative networks, but their exoskeleton and respiratory systems limit their physical size. They cannot grow brains large enough to hold abstract, self-aware reasoning.

The Systemic Reason for the Humanoid Form

From model's perspective, the humanoid template was chosen because it represents the **all-in-one physical package** required for an intelligent species to transform raw thought into physical reality:

[Bipedal Upright Posture] ──► Frees two limbs entirely from walking.
+
[Opposable Precision Hands] ──► Allows fine toolmaking, writing, and engineering.
+
[Dry-Land Environment] ──► Gives access to fire, metallurgy, and chemistry.
+
[Vocal Anatomy & Face] ──► Enables complex speech and rich facial empathy.

1. **Upright Balance (Efficiency):** Standing on two legs frees two dedicated limbs (our arms and hands) entirely for manipulation and creation, rather than walking.
2. **Precision Dexterity:** Opposable thumbs allow us to hold a heavy hammer *and* thread a fine needle, bridging raw physical force with delicate technology.

3. **Dry-Land Mobility:** Living on land allows us to harness fire—the foundational chemical trigger needed to smelt metals, generate electricity, and build industrial technology.

The humanoid form is not an arbitrary shape. It is the **optimal physical hardware blueprint**—the precise biological suit that bridges high-level abstract thought with the physical tools needed to build a civilization, exercise moral choice, and understand the universe.

The Cosmic Architecture Perspective: The Ideal Biological Synthesis

While many animal species exhibit intense emotional telemetry—protecting their offspring with fiercely selfless love, enduring extreme wilderness hardship, and forming deep social packs—their bodies lack the physical tools to transform that care into cumulative technology, written law, or planet-scale stewardship.

To bridge abstract thought with physical creation and unbound moral agency, the physical vessel required a specific **three-way architectural synthesis**:

[PHYSICAL VULNERABILITY] ──► (Shared with Animals: Pain, Hunger, Mortality)
+
[DEEP SOCIAL DEPENDENCE] ──► (Shared with Animals: Extended Infant Care & Empathy)
+
[DEXTEROUS CREATIVE TOOLS] ──► (Unique Synthesis: Opposable Hands & Land-Based Fire)
│
▼
[THE HUMANOID BLUEPRINT] ──► Complete Vehicle for Moral Agency & Technology

1. **Physical Vulnerability (Shared Baseline):** Like the animal kingdom, the human body had to feel hunger, cold, fatigue, and

physical mortality. If an organism never experiences hardship or physical limits, its choices carry no real systemic weight or consequence.

2. **Extended Social Dependence (Nurturing Empathy):** Building on the emotional telemetry perfected across the animal kingdom, human infants are born exceptionally helpless due to rapid brain expansion. This forced long years of parental bonding, community cooperation, and social empathy, turning instinctual care into abstract moral frameworks.
3. **Physical Capability (Hands & Tools to Execute):** This is where the humanoid form diverged. An intelligent, deeply loving mind trapped inside a body without hands (like a dolphin or whale) can feel profound empathy, but it can never smelt metals, write down mathematical laws, build medical tools, or shape its environment.

The Key Insight: The human body did not invent love or hardship. Rather, it served as the **complete hardware vessel** that merged the deep emotional telemetry and vulnerability of the animal world with the physical dexterity needed to turn moral choices into civilization, culture, and planetary stewardship

2. The Terminal Hardware Ceiling: Built-In Limits of the Brain

The incredible success of human civilization often leads us to assume that human intelligence can eventually understand everything in existence. We build powerful supercomputers, split the atom, and program artificial intelligence. It is easy to fall into the trap of believing that there are no ultimate limits to what our minds can grasp. However, every physical hardware instrument ever built has inherent processing boundaries:

```
+--------------------------------------------------------------------------+
| THE 3D SANDBOX HARDWARE LIMITS                                           |
+---------------------+----------------------------------------------------+
| 1. Sensory Bandwidth | Eyes detect <1% of light; ears hear limited acoustic
ranges.                                                       |
| 2. Dimensional Lock | Hardwired to perceive and model only 3 spatial di-
mensions.                                                     |
```

| 3. Processing Engine | Bound by biological neural transmission speeds and chemistry. |
+--------------------+--+

In the language of the *Cosmic Architecture*, the human brain is a **localized 3D terminal**. It evolved specifically to help biological organisms hunt, navigate 3D physical space, and communicate in small groups.

Just as a chimpanzee's biological brain lacks the neural architecture required to comprehend quantum calculus—no matter how patiently you try to teach it—the human brain possesses strict hardware ceilings regarding higher-dimensional geometry, multiverse state-spaces, and the meta-laws of ultimate reality.

The limit is not a lack of curiosity or effort; it is a **bandwidth limitation of local biological hardware**.

3. Emotions as the User Interface: Why Intelligence Needs Feeling

A common misconception is that intelligence alone builds civilization, and that emotions are primitive weaknesses that interfere with logic.

Nature proves the exact opposite. Pure logic calculates what is possible, but **emotion determines what has value**. Imagine a civilization composed entirely of unfeeling, hyper-intelligent biological robots:

- Without **love and bonding**, parents would not endure years of sacrifice to raise helpless infants.
- Without **empathy and trust**, large-scale trade, justice systems, and communities would collapse.
- Without **curiosity**, scientific exploration would never begin.

Emotions act as the **subjective user interface (UI)** of human consciousness. They compress trillions of complex background biological variables into unified, felt experiences. Love, grief, loyalty, and compassion are the essential software patches that enable moral agency, karma, and conscious stewardship in a physical sandbox.

4. Over-the-Air Infusions & The Prepared Mind

(**Note:** As explored in detail in **Chapter 20** regarding Gifted Minds, individual breakthroughs and prodigies reflect specialized access permissions behind the Knowledge Firewall.)

When viewed across the whole species, when civilization reaches a critical evolutionary milestone, within the model, we can imagine such guidance as analogous to an over-the-air knowledge infusion.

Sudden, world-changing leaps in human history—such as the simultaneous discovery of calculus by independent thinkers on opposite sides of the world, or sudden revolutions in ethical thinking—occur when local biological terminals align with higher system updates delivered across the cosmic network.

5. Beyond the Human Stage: Future Intelligence

Humanity is not the final product of creation; we are an active intermediate chapter.

Just as early single-celled organisms could never have imagined a human mind contemplating the stars, humanity today cannot fully visualize the forms of consciousness that will evolve over the next million or billion years.

As our species learn to govern its technical power with moral restraint, the **Cosmic Knowledge Firewall** gradually lowers. Future intelligence will not merely mean faster calculation or bigger computers; it will mean the expansion of moral wisdom, higher-dimensional awareness, and direct participation in the master network of the cosmos.

Chapter 43

Why Earth Became a Water World

Water looks deceptively ordinary because it surrounds us, fills our atmosphere, and composes most of our biological bodies. Yet, a planet capable of maintaining stable, long-lasting surface liquid water over billions of years requires an extraordinary and highly specific convergence of cosmic, geological, and thermodynamic parameters.

Water ice is abundant across the universe, but liquid water on a planetary surface is a rare and delicate equilibrium. Understanding how Earth became a water world requires examining both the physical mechanisms of planetary formation and the deeper architectural purpose of liquid water within a life-bearing system.

1. Thermodynamic Stability: The Three Pillars of a Water World

Liquid water ice is common across deep space, but stable, long-lasting surface oceans are a delicate cosmic achievement. For a planet to maintain liquid oceans over billions of years without drying out or freezing solid, it must satisfy three foundational physical pillars:

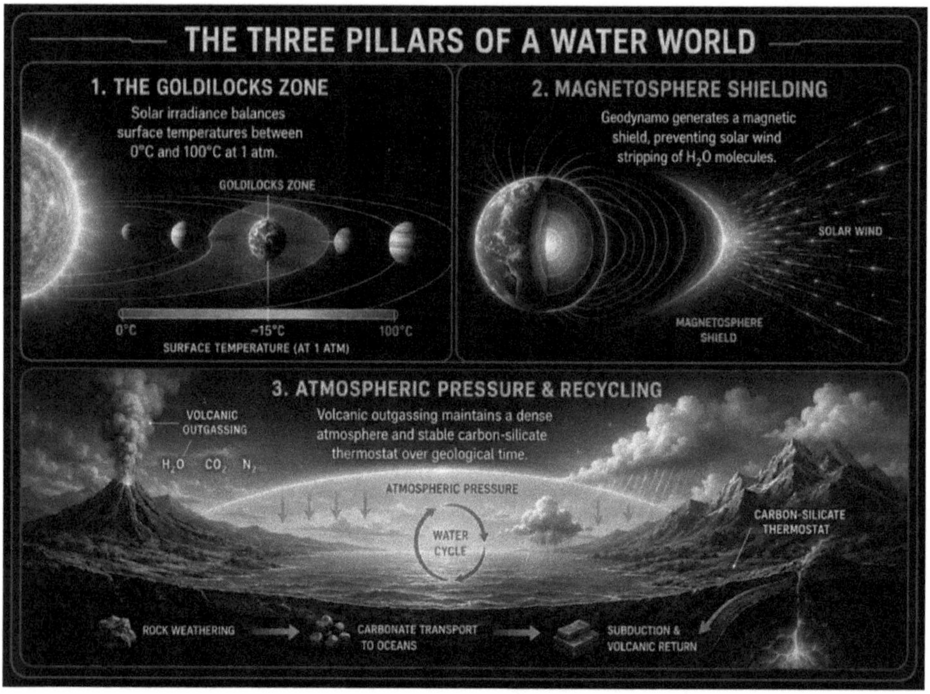

1. The Goldilocks Orbit (Circumstellar Habitable Zone)

Earth orbits at a sweet spot roughly 150 million kilometres from our Sun (a distance astronomers call 1.0 AU). At this precise location, sunlight delivers just enough radiative heat to keep surface temperatures in a gentle window between and.

- **Avoiding the Freeze:** If Earth were even a little farther out, water would freeze into permanent global ice sheets, locking nutrients in cold storage.
- **Avoiding the Furnace:** If Earth were slightly closer to the Sun, solar radiation would trigger a **Runaway Greenhouse Effect**—an overheating trap where surface heat boils the oceans away into atmospheric steam. Once trapped in the upper atmosphere, sunlight splits those steam molecules into hydrogen and oxygen,

letting the light hydrogen escape into deep space forever. This catastrophic breakdown turned Earth's twin sister, Venus, into a bone-dry, toxic furnace.

2. The Magnetic Force Field (The Geodynamo)

Liquid water cannot survive on a planet's surface without a protective blanket of air, and an atmosphere cannot survive without magnetic protection.

Deep inside Earth, continuous heat drives the churning of liquid molten iron in our outer core. This internal metallic generator—known as the **Geodynamo**—creates a massive, invisible magnetic shield (the **Magnetosphere**) that extends far out into space.

- **Stopping Solar Wind Stripping:** The Sun constantly blasts out a fierce stream of high-energy, charged particles called the solar wind. Earth's magnetic force field acts as a planetary deflector shield, bouncing these high-speed particles away.
- **The Lesson of Mars:** Mars was once a warm world with rivers and surface lakes. However, because Mars is smaller than Earth, its internal core cooled down, causing its magnetic generator to fail. Without a magnetic deflector shield, the relentless solar wind stripped away the Martian atmosphere over millions of years, causing its oceans to boil off into the vacuum of space.

3. Air Pressure and Volcanic Climate Recycling

Even with the right temperature and a magnetic shield, liquid water cannot exist in a vacuum. It requires the physical weight of an atmosphere—**Atmospheric Pressure**—pressing down on the surface to hold water molecules together in liquid form. Without enough air pressure, surface water instantly boils away into gas, regardless of how cold it is.

Earth maintains this vital air pressure through continuous geological recycling:

- **Volcanic Outgassing:** Volcanoes act as planetary pressure valves, spewing steam, carbon dioxide, and nitrogen out from deep within Earth's mantle to continuously replenish our atmospheric blanket.
- **The Carbon-Silicate Thermostat:** Earth features a long-term "planetary air conditioner". Rain absorbs atmospheric carbon dioxide and carries it down into ocean rocks. Over geological epochs, tectonic plates slide beneath one another (**subduction**), dragging those carbon-rich rocks deep into Earth's mantle where they are heated and eventually vented back out by volcanoes as gas. This automated thermostat keeps planetary greenhouse gas levels balanced, preventing Earth from sliding into an endless ice age or an overheated furnace.

2. Dual Origins: Water from Inside and Water from Above

Where did Earth's massive volume of water—roughly 1.38 billion cubic kilometres—come from? Geochemical evidence demonstrates that Earth's oceans were built through a two-source hybrid engine: internal planet creation combined with external deliveries from space.

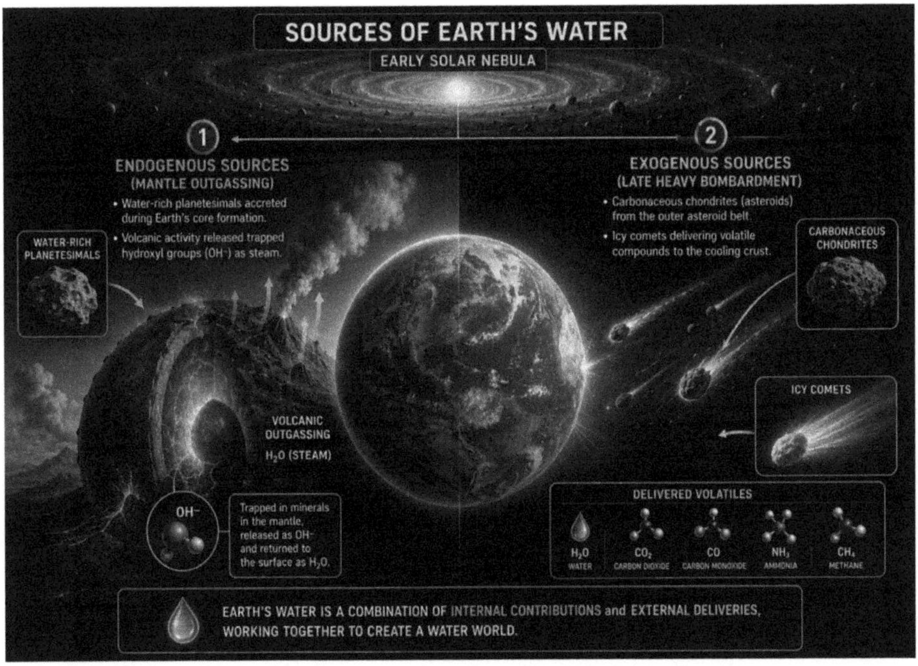

1. Internal Origins (Mantle Outgassing: Water from Below)

When Earth was first forming from spinning clouds of space dust and colliding space rocks (**planetesimals**), water ingredients were incorporated directly into the young planet's deep rocky layers.

- **Water Locked in Rock:** Inside Earth's deep mantle, water does not sit as underground liquid pools. Instead, it exists as chemical building blocks—**trapped hydroxyl groups** (OH^-)—bound tightly inside solid silicate minerals.
- **Volcanic Steam Releases:** As Earth's core heated up from radioactive decay and intense gravitational pressure, these mantle minerals melted. The trapped hydroxyl building blocks recombined into water steam and were blasted up to the surface through massive volcanic plumes. As the young planet's crust cooled, this

volcanic steam condensed into continuous rains that filled Earth's primordial ocean basins.

2. External Deliveries (The Late Heavy Bombardment: Water from Space)

While internal volcanic venting provided a huge portion of Earth's water, it was not the whole story. Early in Solar System history, Earth experienced heavy impact activity from asteroids and other small bodies. A sharply defined Late Heavy Bombardment around roughly 4.0 to 3.8 billion years ago has been proposed, although its timing and intensity remain subjects of scientific debate.

- **Space Rocks and Dirty Snowballs:** Millions of carbon-rich asteroids (**carbonaceous chondrites**) from the outer asteroid belt and icy comets from the frozen fringes of space slammed into the cooling Earth.
- **Delivering Volatiles:** These impacting space bodies acted as interstellar delivery trucks, dropping off vast payloads of **volatile compounds**—easy-to-evaporate liquids and gases such as water ice (H_2O), carbon dioxide (CO_2), methane (CH_4), and ammonia (NH_3).
- **The Chemical Fingerprint:** Scientists proved this space delivery happened by measuring the chemical fingerprint of ocean water (specifically the ratio of heavy hydrogen, or deuterium, to normal hydrogen). The chemical fingerprint of Earth's oceans matches the exact chemical fingerprint of carbonaceous chondrite asteroids, provides strong evidence that water-rich asteroidal material contributed substantially to Earth's water inventory.

3. Solar System Comparisons: Why Earth Is Unique

Earth is far from the only body in our Solar System containing water, but it is unique in *where* and *how* that water is maintained:

Body	**Water State**	**Primary Location**	**Surface Life Suitability**
Earth	**Liquid / Vapor / Ice**	**Surface Oceans & Atmosphere**	**Optimal (Stable 3D Environment)**
Europa (Jupiter Moon)	Liquid Ocean	Subsurface (under tens of km of ice)	Limited (No direct solar energy / Deep Ocean)
Enceladus (Saturn Moon)	Subsurface Ocean / Geysers	Interior crust	Limited (Geothermal vents only)
Titan (Saturn Moon)	Liquid Hydrocarbons / Water Ice	Surface Methane Seas / Ice Crust	Alternative chemistry (non-water based)
Mars	Polar Ice / Subsurface Deposits	Crust / Atmosphere	Extinct surface oceans (Atmosphere stripped)

While sub-surface oceans on Europa or Enceladus may harbor microbial life around deep-sea hydrothermal vents, surface liquid water on Earth provides something far more vital for complex evolution: **direct access to solar radiation (photosynthesis), seasonal variation, land-water interfaces, and a dynamic atmospheric cycle.**

4. The Cosmic Architecture Perspective: Water as the Universal Solvent

In the *Cosmic Architecture*, water is treated not as a mystical substance but as an unusually powerful physical medium. It stabilises climate, dissolves and transports molecules, supports cellular chemistry and creates a broad environment in which biological complexity can develop.

- **The High Dipole Moment & Molecular Transport:** is a polar molecule with an exceptional ability to dissolve salts, minerals, and organic compounds. In a biological context, water acts as the ultimate transport medium, allowing amino acids, nucleic acids, and metabolic nutrients to move, interact, and self-assemble into complex cell structures.
- **Thermal Regulation & Planetary Sandbox Stability:** Water possesses an unusually high specific heat capacity (~4,184 J/kg·°C). Earth's vast oceans absorb massive amounts of solar radiation during the day and release it slowly at night, buffering global climate extremes. Without Earth's oceans acting as enormous thermal reservoirs, surface temperature variations would be substantially more extreme, creating far fewer stable conditions for the evolution of complex life as we know it.
- **The Matrix for Biological Deployment:** Before land plants or complex air-breathing animals could be deployed, life required a three-dimensional fluid environment where buoyancy reduced gravitational constraints, allowing delicate organic structures (protocells, RNA chains, and early marine organisms) to evolve without collapsing.

Earth became a water world because **water is the ideal physical hardware substrate for organic complexity.** The laws of physics, nuclear fusion inside stars, planetesimal accretion, and planetary magnetic dynamics were pre-configured to produce a world where water could remain liquid across geological epochs—creating the ultimate physical stage for biological agency to emerge.

PART X — KARMA, HIERARCHY AND INTERVENTION

Religious concepts examined without reducing them to software

Looking Back:

In **Part IX**, we synthesized the overarching blueprint of reality into a single, unified view. We reflected on how systems philosophy allows us to bridge science and spirituality with intellectual discipline, wonder, and moral clarity.

Moving Forward:

In this final, specialized section, **Part X** explores deep, specific edge-cases within our framework. We dive into the profound mechanics of cross-layer intervention—asking exactly why and under what system conditions an Avatar enters a physical sandbox. We examine the Jain concept of *Kevala Jnana* (unbounded, real-time omniscient state access) through the lens of root-level system clearance. Finally, we return to the Cosmic Boundary—contemplating why the system isolates our current view, and why our ultimate purpose is not to escape the sandbox, but to master moral agency and stewardship right where we stand.

Chapter 44

Why Would an Avatar Enter the System?

While Chapter 27 detailed the systems mechanics of an Avatar as a root-level patch, a deeper philosophical question remains: *Why choose the slow, costly, and ambiguous route of living inside human history rather than transmitting raw data from above?*

A person can be told what is right and still fail to live differently. Information alone does not always transform behaviour. A teacher, reformer or religious figure may influence people by living among them, forming relationships and acting within history. The *Cosmic Architecture* uses this to ask why a higher-level intervention might take an embodied form rather than simply sending information from outside.

Information is not transformation

People often know what they should do and fail to do it. Moral change is not simply a data-transfer problem. A perfect instruction manual cannot guarantee courage, compassion or restraint. If the purpose of an avatar is transformation rather than information, embodiment becomes more intelligible.

A teacher can create communities, rituals, narratives and examples that continue after death. These social processes can alter a civilisation for centuries. The architecture so should not imagine intervention as uploading facts into brains. It may be closer to seeding a durable cultural attractor. This way of understanding it remains compatible with purely human explanations of religious founders. Its philosophical or spiritual significance depends on additional premises.

Working through ordinary causality

A cross-layer intervention need not resemble a violation of physics. If a higher-level cause acts through an embodied agent, the visible chain may consist of ordinary speech, travel, political conflict and social influence. This avoids using "miracle" as a placeholder process.

But it raises an evidential problem: *if all observable effects are ordinary, how could one distinguish divine intervention from unusual human*

agency? Theology may appeal to revelation or tradition; science would require publicly testable differences.
The architecture should admit that this distinction may not be observational or experimentally decidable.

Freedom and non-coercion

An embodied teacher can be accepted, rejected, misunderstood or opposed. That ambiguity preserves room for human response. An irresistible supernatural display would produce a different moral environment.
This offers one philosophical reason for subtle intervention: the aim may be participation rather than control. Yet this is a theodicy-like explanation, not evidence that such intervention occurs.
The strongest version of the chapter so presents embodiment as logically consistent under certain theological assumptions while keeping historical truth-claims separate.

Repair without erasure

The local-intervention model connects avatar theory to cosmic maintenance. A civilisation contains accumulated language, art, relationships and knowledge. If disorder can be corrected without annihilating that inheritance, preservation appears rational.
Within the *Cosmic Architecture*, history is not treated as disposable. The model prefers repair, redirection and renewal to repeated destruction. An avatar or higher intervention could correct a damaged local section without resetting the whole universe, just as a complex system can be repaired without deleting everything it has learned.

Chapter 45

Kevala Jnana and the Possibility of Unbounded Knowledge

Every human being has experienced the feeling of not knowing something. A child cannot understand advanced mathematics. A medical student cannot perform surgery on the first day of university. Even the greatest scientists freely admit that there are questions they cannot yet answer. Human knowledge has always been limited, and perhaps recognising those limits is one of the reasons civilisations continues to progress.

Yet throughout history, many spiritual traditions have described individuals who appeared to possess an extraordinary level of understanding. They were called enlightened, awakened, liberated or omniscient. ***Jain*** philosophy offers one of the most detailed descriptions through the concept of **Kevala Jnana**, often translated as perfect or unlimited knowledge. Among all the ideas explored in this book, this may be one of the most profound because it asks a question that science itself has not yet answered: **Can consciousness develop beyond the ordinary limits of the human mind?**

Before exploring this question through the *Cosmic Architecture*, it is important to understand what Jain philosophy teaches. Kevala Jnana is not a reward randomly granted by God, nor is it a magical power that appears without preparation. Jain philosophy describes an extraordinarily long journey in which the soul gradually removes the karmic conditions that obscure its natural capacity for knowledge. This journey is measured not in years but potentially across countless lifetimes. Progress is achieved through ethical living, self-discipline, non-violence, truthfulness, detachment and spiritual purification. Only when these obscuring conditions have been completely removed does Kevala Jnana arise.

Whether one accepts this teaching literally or symbolically, the underlying idea is remarkably thought-provoking. Human consciousness may not be fixed. It may continue to develop.

Modern science already accepts a simpler version of this principle. Children gradually become capable of understanding ideas that were once impossible for them. Education transforms the way people think. Experience deepens judgement. Neuroscience shows that the brain continually reorganises itself throughout life, forming new neural connections as we learn. Although science does not describe this process as spiritual awakening, it clearly demonstrates that human understanding is not static.

The *Cosmic Architecture* begins with this same observation but extends the question much further. *What if today's human brain represents only one stage in the long evolution of consciousness?*

We often assume that if something exists, humanity will eventually discover it. History encourages this confidence because our knowledge has expanded dramatically over the past few centuries. Yet every measuring instrument has limits. Every species has limits. Even our senses detect only a tiny fraction of reality. Human eyes cannot see radio waves or X-rays without specialised instruments. Human ears perceive only a narrow range of sound frequencies. Our memories are imperfect, and our reasoning is influenced by emotion, experience and cognitive bias.

Why should the human brain itself be the only thing without limits?

Perhaps the greatest limitation facing humanity is not a lack of effort but the architecture of the mind itself. Imagine trying to explain quantum mechanics to a chimpanzee. No amount of patience could overcome the biological limitations of its brain. The problem would not be intelligence in a general sense but cognitive architecture. The brain simply lacks the structures required to understand such highly abstract ideas.

If every other species has cognitive limits, why should *Homo sapiens* assume that we have reached the final form of intelligence? Perhaps we stand where earlier hominids once stood. They could never have imagined relativity, artificial intelligence or space exploration. Likewise, humanity today may be unable even to imagine levels of understanding that future forms of consciousness will one day consider ordinary.

This possibility changes how we think about Kevala Jnana. Instead of viewing it as a supernatural miracle, we might understand it as describing an extraordinary transition in consciousness itself. The language differs between science and spirituality, but both recognise that understanding can expand. Jain philosophy simply extends that development to its conclusion.

The *Cosmic Architecture* proposes another possibility. Suppose reality is organised by a civilisation billion or even trillions of years older than our own. Such a civilisation would have experienced more history than humanity can comprehend. It may have observed countless biological experiments, civilisations and perhaps even multiple universes. Over such immense periods, it would almost certainly have developed forms of knowledge, governance and security far beyond anything we currently understand.

Human civilisation already protects dangerous knowledge. Nuclear technology is tightly regulated. Biological research operates under strict safety protocols. Military systems require multiple layers of authorisation. These restrictions do not exist because knowledge is evil but because powerful knowledge, given too early or without responsibility, can cause catastrophic harm.

Would a civilisation that has survived for trillions of years behave with less wisdom than humanity?

Probably not.

The *Cosmic Architecture* therefore introduces a philosophical possibility: perhaps knowledge itself is permission-based—not because someone is deliberately hiding it, but because consciousness must first become capable of receiving it.

This pattern appears repeatedly throughout nature. Children learn arithmetic before calculus. Doctors master anatomy before performing surgery. Pilots train in simulators before flying passengers. Every complex skill develops in stages because understanding cannot simply be downloaded into an unprepared mind.

Perhaps reality itself follows the same principle.

If so, Kevala Jnana would represent far more than possessing enormous amounts of information. It would represent the complete removal of the limitations that normally prevent consciousness from perceiving reality directly.

This distinction is important. Modern computers can store vastly more information than any human being, yet they do not possess wisdom. A search engine contains billions of documents but does not experience understanding in the way conscious beings do. Knowledge is not merely data, and wisdom is not merely storage capacity.

For this reason, Kevala Jnana should not be imagined as opening every folder inside a cosmic database. A better analogy is the gradual lifting of a dense fog from a landscape. The mountains were always there. The observer simply could not see them clearly.

The *Cosmic Architecture* extends this analogy one step further. Perhaps every conscious being exists somewhere along a spectrum of awareness. Most of us perceive only a tiny fraction of reality because our present state naturally limits what we can understand. As consciousness develops, those limitations gradually diminish. Whether that development occurs through many lifetimes, biological evolution, moral growth or processes we have not yet discovered remains an open question. The important idea is not that knowledge is artificially withheld but that understanding itself must grow before deeper knowledge becomes meaningful.

This perspective also raises an interesting question about gifted individuals. Every generation seems to produce people whose abilities appear extraordinary. Some children display remarkable mathematical intuition. Others compose music of astonishing complexity while still very young. Science explains much of this through genetics, environment and education, and those explanations remain essential. The *Cosmic Architecture* simply asks whether exceptional minds might also represent small glimpses of what future human consciousness could become. They would not possess unlimited knowledge, but they might occasionally perceive patterns that most of humanity has not yet learned to recognise.

Whether this idea is true remains entirely speculative. No scientific evidence currently demonstrates that gifted individuals possess privileged access to a deeper reality. Nevertheless, history repeatedly reminds us that civilisation often advances because a small number of people perceive possibilities that others overlook.

Kevala Jnana represents something far beyond even these extraordinary minds. It describes the complete removal of limitation itself.

Whether such a condition is literally possible is a question that neither science nor philosophy has yet resolved. Neuroscience has not discovered unlimited consciousness, and Jain philosophy approaches the subject from an entirely different framework. The Cosmic Architecture therefore treats Kevala Jnana neither as established science nor as ancient mythology. Instead, it presents it as one of humanity's most sophisticated attempts to explore perhaps the deepest question of all:

What would reality look like if consciousness no longer had limits?

Perhaps the answer lies beyond today's science.

Perhaps it lies beyond today's philosophy.

Or perhaps it simply lies beyond today's human brain.

Chapter 46

The Cosmic Boundary: Why We May Not See the Whole

Human beings often assume that reality is fully accessible to us. If something exists in nature, we expect that building better telescopes, designing more powerful supercomputers, or discovering more advanced mathematics will eventually reveal it to us.

History constantly encourages this confidence. Scientific instruments have repeatedly expanded the horizons of human observation:

- **Microscopes** revealed a microscopic world of living cells and microorganisms previously unimaginable to ancient thinkers.
- **Telescopes** expanded our reach from local constellations out to billions of distant galaxies.
- **Radio astronomy** uncovered a dynamic universe humming in frequencies completely invisible to human eyes.
- **Particle detectors** mapped subatomic interactions taking place far below the threshold of ordinary sensory experience.

Yet, an ironic pattern runs through the history of discovery: **every single expansion of human knowledge has also revealed new, fundamental limits**.

Biological Senses and Physical Limits

Human sensory organs detect only a tiny slice of physical reality. The human eye perceives less than one percent of the total electromagnetic spectrum. Radio waves, microwaves, infrared radiation, ultraviolet light, X-rays, and gamma rays surround us continuously, yet without instruments, they remain completely imperceptible to our biological senses.

Human hearing is restricted by similar acoustic boundaries. Countless animal species detect magnetic fields, ultrasonic frequencies, and chemical signals that remain entirely outside human perception. While technology undeniably extends our senses, our instruments possess their own hard limits. Every telescope has a sensitivity range, every detector suffers from background noise, and every scientific experiment measure only specific property under controlled conditions.

Human beings, therefore, never observe raw, unfiltered reality. We experience a localized version shaped by our biology, our tools, and the physical perspective from which we observe.

Structural and Spacetime Boundaries

This inherent limitation does not imply that a mysterious entity is actively concealing knowledge from humanity. A boundary can exist naturally without functioning as a deliberate security wall.

In modern physics, the **observable universe** is fundamentally bounded because light travels at a finite speed (c) and the universe is a finite age. We cannot receive physical signals from regions beyond this cosmic horizon simply because the light has not had enough time to reach our position.

Similarly, a **black hole event horizon** represents a purely geometric boundary in spacetime. It is not a spiritual wall or a fence built to keep observers away; it is a structural feature of general relativity that defines which cause-and-effect events can interact with the outside universe.

These scientific examples are crucial because they demonstrate that inaccessible regions are not automatically imaginary. Reality already contains objective boundaries created by geometry, time, and causality. The *Cosmic Architecture* uses these established physical facts as a starting point to ask a broader question: **Could there be deeper boundaries created by higher dimensions, state configurations, or levels of conscious maturity that humanity has not yet mapped?**

The Observer as the Ultimate Boundary

The most significant constraint on human understanding may not be a physical wall located somewhere in deep space. The ultimate boundary may reside inside the observer.

RECEPTOR CAPACITY	**STIMULUS PRESENT**	**RESULTING PERCEPTION**
Canine Mind	Mathematical equations written on a board.	Perceives chalk marks; completely blind to abstract algebraic logic.
Infant Mind	Complex legal agreement or enterprise contract.	Sees ink shapes on paper; cannot decode legal obligations.
Untrained Listener	High-dimensional orchestral symphony.	Hears general sound; misses intricate counterpoint and formal structure.

These everyday human examples highlight a vital distinction: **seeing and knowing are not the same thing.**

A radio signal can travel through a room without anyone noticing its presence. The signal only becomes meaningful sound when a compatible receiver tunes into the frequency and decodes the incoming transmission. A written language remains mere scribble on paper until an educated mind decodes its grammar and vocabulary. In every domain of existence, information may be present everywhere, yet meaningful access remains strictly bounded by the internal capacity of the receiving system.

The Dimensional Firewall and Conscious Scope

In systems engineering, access control is not designed to hide information out of spite; it ensures that a process only receives data it has the framework to handle safely.

If humanity currently operates inside a biological, 3D experimental sandbox, our sensory equipment and linear-time cognition are tuned specifically for physical survival. We are optimized to dodge physical hazards, build local tools, and navigate linear time. Asking a biological brain to directly perceive higher-dimensional state-space or the underlying mechanics of the *Cosmic Knowledge Firewall* is like expecting a simple transistor radio to render complex 3D virtual reality.

The boundary around our perception is both a physical reality and a developmental safeguard. It ensures that humanity learns to master its immediate physical world—building moral integrity, environmental stewardship, and social balance—before gaining access to higher operational tiers of the cosmic framework.

Question to Consider
If the information defining higher layers of reality is already passing through the space around us, what internal adjustments in human awareness, moral maturity, and scientific perspective will be required for us to build the receiver?

PART XI – THE DEEPER COSMIC FRAMEWORK

Having explored the final questions of karma, hierarchy, intervention, knowledge and cosmic limitation, we can now step back from the individual components and view the architecture as a whole. This final Part does not introduce another major mechanism. Instead, it gathers the book's central ideas into a single integrated view before closing with a reflection on wonder, evidence and intellectual discipline.

The Architecture in One View

This final chapter gathers the book's central concepts into a single view: systems, memory, state-space, consciousness, hierarchy, non-interference, cosmic inheritance and disciplined wonder. It does not introduce a new doctrine; it shows how the earlier chapters connect.

Afterword – Wonder with Discipline

Wonder opens the door to large questions, but discipline prevents wonder from becoming confusion. The Cosmic Architecture is offered in that spirit: bold enough to imagine a deeper reality, careful enough to respect science, and humble enough to admit where speculation begins.

The Cosmic Architecture began with intuitions that many readers share: the universe seems astonishingly structured; consciousness feels difficult to reduce to equations; religious traditions preserve ambitious claims about moral order and levels of reality; and modern information systems provide new metaphors for relation, memory and hierarchy. The expanded edition has tried to keep the wonder while removing the shortcuts.

The central lesson is that explanation has layers. Learning how planets form does not make their existence less remarkable. Understanding evolution does not eliminate the philosophical question of why a universe capable of life exists. Explaining a sudden insight through neural processing does not settle the nature of consciousness. At the same time, none of these remaining questions licenses us to declare a preferred metaphysics proven.

The architecture is offered as a map of possibilities and as the author's proposed model of a deeper reality. Its nodes represent actual beings; its vectors represent ways those beings might be related; its sections represent larger communities or realms; its hierarchy represents different levels of knowledge, responsibility and access; and its cosmic memory represents the possibility that valuable history is preserved. Human technology supplies the metaphors, but the proposed cosmic system would be vastly more advanced than anything humanity has built.

If future evidence supports some part of the architecture, the model should become more precise. If evidence contradicts it, the model should change. If some questions remain permanently philosophical, they should be recognised as such. A worldview becomes stronger when it can survive correction.

This revised edition also preserves a second discipline: correction should not erase the originating idea. The Cosmic Architecture is allowed to make a bold proposal—actual conscious beings in a deeper relational state-space, rare transformations of access, differentiated higher-scope agency, preservation of accumulated intelligence and possible cosmic inheritance—provided that each proposal is labelled as speculation and distinguished from the religious traditions that inspired parts of its vocabulary.

The result is deliberately neither a theology textbook nor a disguised claim of new physics. It is a philosophical model in which Jain, Hindu, Buddhist and Christian concepts can challenge one another; systems engineering can provide structural analogies; and science can set hard constraints on what may responsibly be claimed.

Wonder and discipline are not enemies. Wonder supplies the courage to ask enormous questions. Discipline prevents enormous questions from receiving easy answers. Between them lies the possibility of a mature cosmic philosophy: one that respects what humanity has learned, admits what it has not learned, and continues asking what kind of reality could contain matter, life, mind, value and the desire to understand them all.

Perhaps the greatest product of cosmic evolution is not life itself, nor intelligence, nor technology. Perhaps it is the gradual emergence of beings capable of understanding the architecture from which they arose—and, in doing so, becoming conscious participants in its continuing evolution.

www.ingramcontent.com/pod-product-compliance
Lightning Source LLC
LaVergne TN
LVHW051921051126
851857LV00021B/2094